KU-166-245

Mapping the Planets

Mapping the Planets

Discovering the worlds beyond our own

Anne Rooney

This edition published in 2019 by Arcturus Publishing Limited
26/27 Bickels Yard, 151–153 Bermondsey Street,
London SE1 3HA

ISBN: 978-1-78888-735-9
AD006699UK

Printed in Singapore

CONTENTS

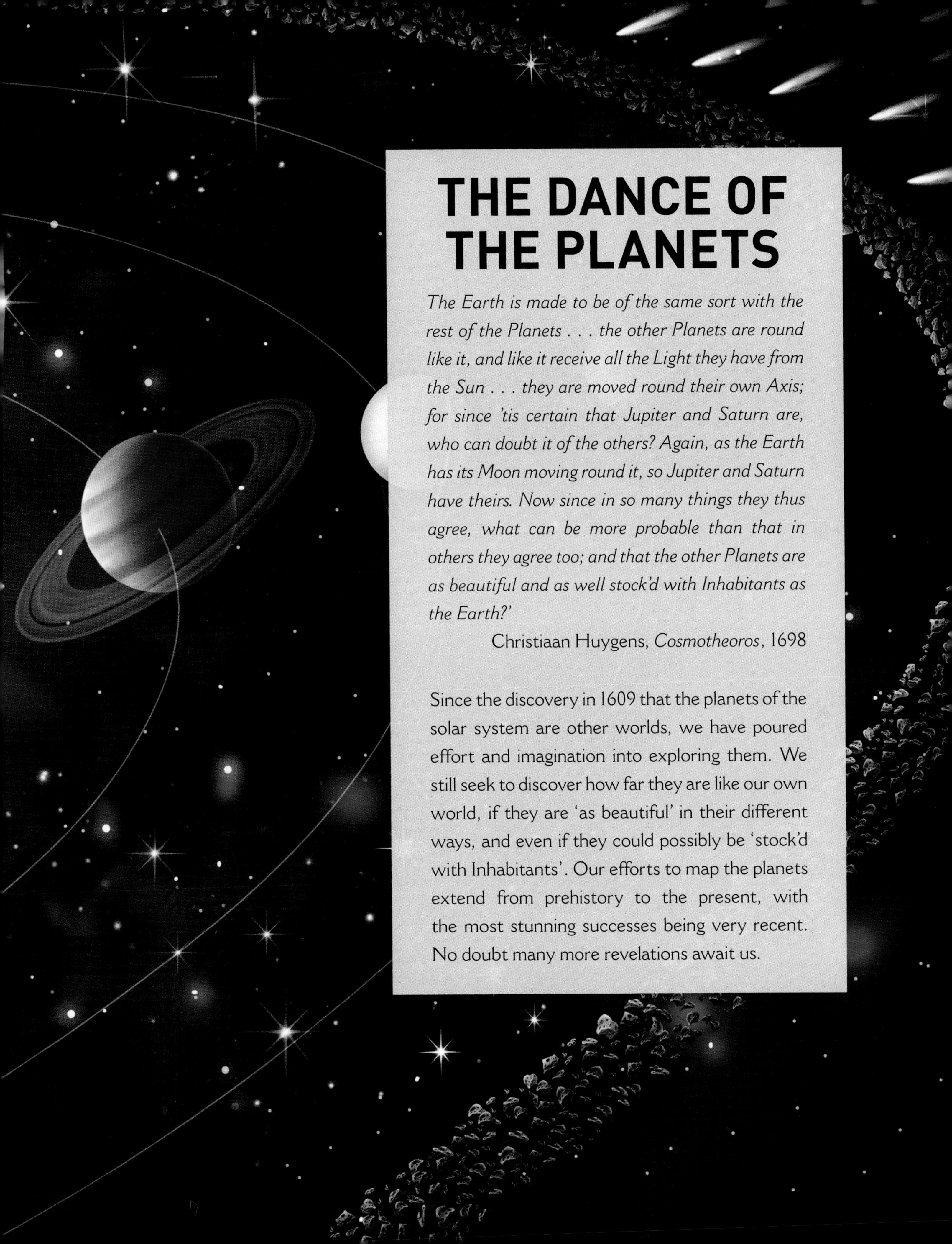

THE DANCE OF THE PLANETS

The Earth is made to be of the same sort with the rest of the Planets . . . the other Planets are round like it, and like it receive all the Light they have from the Sun . . . they are moved round their own Axis; for since 'tis certain that Jupiter and Saturn are, who can doubt it of the others? Again, as the Earth has its Moon moving round it, so Jupiter and Saturn have theirs. Now since in so many things they thus agree, what can be more probable than that in others they agree too; and that the other Planets are as beautiful and as well stock'd with Inhabitants as the Earth?'

Christiaan Huygens, *Cosmotheoros*, 1698

Since the discovery in 1609 that the planets of the solar system are other worlds, we have poured effort and imagination into exploring them. We still seek to discover how far they are like our own world, if they are 'as beautiful' in their different ways, and even if they could possibly be 'stock'd with Inhabitants'. Our efforts to map the planets extend from prehistory to the present, with the most stunning successes being very recent. No doubt many more revelations await us.

LOOKING UPWARDS

Our ancestors were clearly astronomers – their art and monuments tell us so, although we don't know how they explained what they saw. They must have noticed that most spots of light twinkle and stay still with respect to one another, making patterns that can be seen reliably night after night and year after year. They would have seen, too, among the thousands of stars, five spots of light that do not twinkle but shine steadily and move through the background stars. The Ancient Greeks would later give these moving lights the name *planetos*, or 'wanderers'. Besides the Sun, they are the principal bodies of our solar system, Earth's siblings. We now recognize eight planets, but only five (besides Earth) are visible to the naked eye and identifiable as planets: Mercury, Venus, Mars, Jupiter and Saturn.

DRAWN TO THE CENTRE

It's an immense step from sighting the planets to building an intellectual model of what they are, studying how they move and how they relate to Earth. This can be done by tracking the moving dots, night after night, and counting the number of days each one takes to return to its first observed position. Indeed, the earliest astronomical records are more than 3,000 years old. Babylonian clay tablets made between 1595 BC and 1157 BC list the first and last risings of Venus over a course of 21 years.

By the 8th century BC, Babylonian astronomers had become so expert at tracking and calculating the positions of the planets for astrological predictions that their observations would be useful to later astronomers The Ancient Greeks are believed to be the first to have modelled the solar system. They had two competing versions: the first one put the Earth at the centre of a system of orbiting bodies; the second put the Sun at the centre. Working simply from observation, it's impossible to tell which of these is correct – both models would look the same from the Earth.

This Chinese oracle bone describes a lunar eclipse which took place on 27 December 1192 BC. The earliest surviving oracle bone records a solar eclipse in 1302 BC.

The Milky Way is a band of billions of stars, visible here over Mo Hin Khao rock formation in Thailand as a cloudy streak. Difficult to see with modern light pollution, it would have been vivid to our ancestors on a clear night.

The most obvious, Earth-centred (or geocentric) model was promoted by the influential Greek philosopher Aristotle in the 4th century BC and further developed by the Egyptian astronomer Ptolemy in the 2nd century AD. This system put the Moon and Sun closest to Earth with the other planets in sequence beyond them and the 'fixed stars' occupying the outermost region. The sequence of the planets could be worked out from the length of time each took to return to an original position, with the shortest time corresponding to the planet with the shortest orbit, which was therefore believed to be the closest to Earth.

The geocentric model matched everyday experience (the Sun seems to move across the sky) and later suited the Judaeo-Christian belief that God had created Earth as a special place for humans to inhabit. At the same time, the model put Earth, with its sinful human inhabitants, far from the encompassing sphere of heaven beyond the circling planets. Situating Hell within Earth perfected this, putting it as far from Heaven as possible. The geocentric model was accepted with little challenge until the middle of the 16th century.

The Sun-centred (heliocentric) model, proposed by Aristarchus of Samos in the 3rd century BC, had nothing to recommend it above the Ptolemaic system and had the drawback of making humans rather less special. It languished without a champion until 1543, when the Polish astronomer Nicolaus Copernicus (Mikołaj Kopernik) proposed it anew. This model was unpopular with the Church as it apparently contradicted the biblical account in the Book of Joshua which describes God stopping the progress of the

Sun across the sky. The heliocentric model was tolerated for a while, though, largely because it was useful for helping to predict planetary movements; it was not regarded as a literal account of the state of the universe.

LOOPING PLANETS

The motions of the planets as seen from Earth are not straightforward.

The progress of each is punctuated by periods of apparently retrograde (reverse) movement, when the planet seems to stall briefly, then goes backwards in its course, before stopping once more and then going onwards again. This pattern, which is the result of us watching from a planet that is itself orbiting the Sun, could not be properly explained until the 1600s.

To accommodate the pattern and predict positions required an elegant workaround that had the planets performing a complex dance in the sky. The workaround was a system of epicycles. Each planet was thought of as going around in a small circle (the epicycle), with the centre of that circle orbiting Earth or a point near Earth in a circular path.

Copernicus' model offered no improvement in predictions as he assumed perfectly circular orbits around the Sun. He still needed epicycles to calculate the correct positions, so there was no particularly compelling reason for astronomers to accept his model over the established Ptolemaic model.

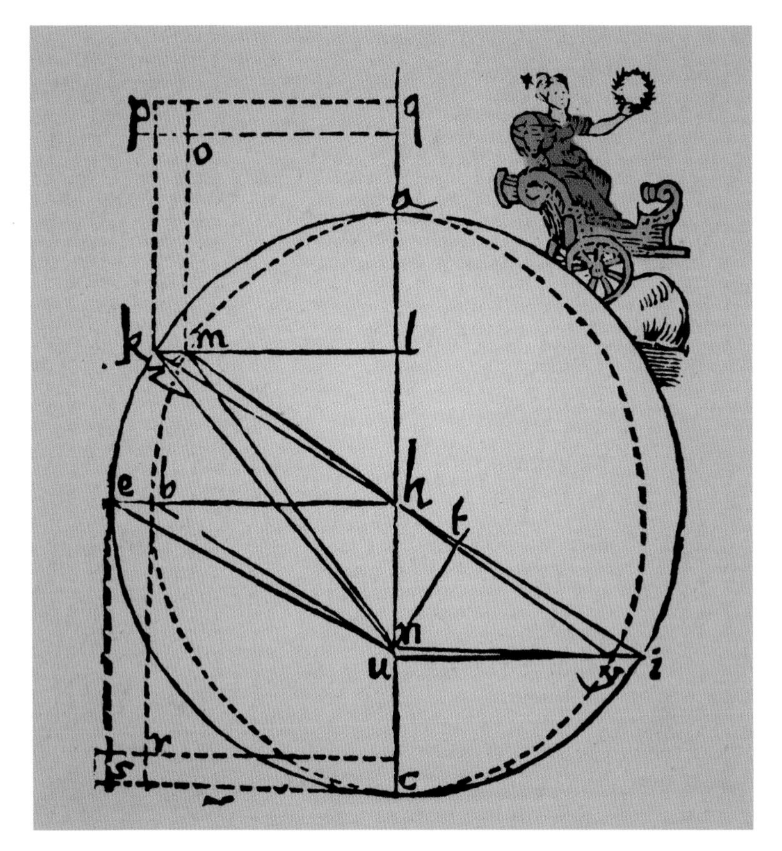

The orbit of Mars as plotted by Johannes Kepler in Astronomia Nova, *1609.*

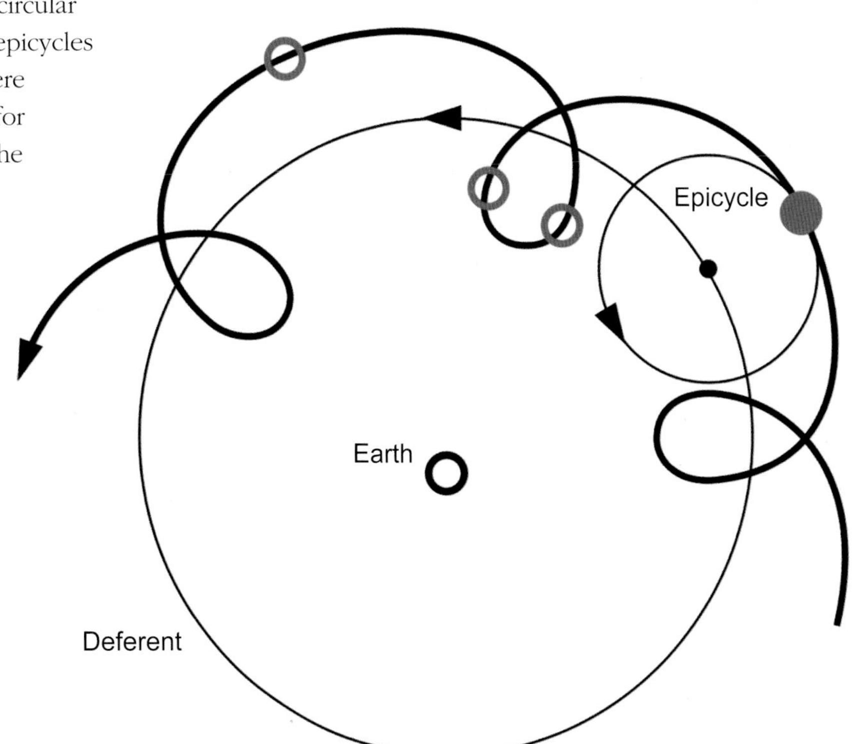

Facing page: *The Ptolemaic model of the solar system put the planets, Moon and Sun in concentric orbits around a central Earth.*

Right: *The apparent movement of a planet (red) in its epicycle on a path around Earth (white).*

SQUASHING THE CIRCLE

Then, in 1609, the talented German astronomer Johannes Kepler used detailed measurements of the orbit of Mars to refine Copernicus' model. He found that the orbit of the planets is not circular but elliptical. No jiggery-pokery with epicycles is needed to explain their intermittent retrograde motion: the apparent loops in their orbits are simply a result of watching from Earth, combining our elliptical orbit with theirs. Under Kepler's system, the positions of the planets could be accurately predicted and all anomalies accounted for. It no longer looked as though the heliocentric model was entirely theoretical, and tensions rose. Another event, at almost exactly the same time, further complicated a delicate situation.

OPENING THE HEAVENS

In 1608, a Dutch lens maker, probably Hans Lippershey, invented the telescope. The Italian scientist Galileo Galilei almost immediately improved it and turned it towards the heavens. His telescope resolved the bright dots of the planets into discs, revealed the cratered surface of the Moon, and showed the Milky Way to comprise a myriad stars. The heavens were no longer how they had always seemed to be, and the world would never be the same again.

Just accepting the existence of the planets as worlds, although troubling for the conventional religious view, was not too dire. But when Galileo discovered moons orbiting Saturn, the geocentric model took a hard knock. If moons could orbit another planet, then it followed that everything in the solar system did not orbit Earth.

Galileo was drawn to and later promoted the heliocentric model, not just as a mathematical convenience but as a literal representation of the heavens. The Church, needless to say, was not impressed. By 1633, Galileo had been accused of heresy and his books and teaching of the heliocentric system were banned. It took more than 300 years for the Catholic Church to approach anything like an apology for their treatment of Galileo and their condemnation of what turned out to be the truth.

Facing page: *Galileo demonstrates his telescope to the Doge of Venice.*

Left: *Copernicus put the Sun at the centre of the solar system, with all the planets in orbit around it and only the Moon still orbiting Earth.*

The Hubble Space Telescope in orbit 568 km (353 miles) above the Earth.

SURFACES IN FOCUS

The invention of the telescope soon made a different type of mapping possible. Instead of just working out the sequence and orbits of the planets, astronomers could begin to look at them as bodies in more detail. Rapid advances in telescope technology brought more and more of the planets' details – and their moons – into view. After 150 years, the telescope also revealed new planets and moons and other smaller bodies, such as dwarf planets and asteroids.

The development of telescopes set high on mountains cut interference from Earth's atmosphere to a minimum; however, a telescope outside Earth's atmosphere is completely unaffected by atmospheric distortion and can gain a clearer view. The Hubble Space Telescope, launched in 1990, soon revealed the planets and their moons in greater detail than ever before. By imaging the bodies in ultraviolet and infrared as well as visible light, Hubble could show up thermal differences and other details not previously seen.

TAKING LIGHT APART

Optical telescopy could go only so far in uncovering the mysteries of the planets. The development of spectroscopy in the 19th century gave astronomers a new and invaluable tool. By examining the light that comes from planets, it's possible to work out which chemical elements are present in them. This is because elements absorb and reflect different parts of the spectrum of visible light, infrared and ultraviolet. Light from a planet is reflected from the Sun, so we can draw conclusions about a planet's composition by comparing its spectrum with that of sunlight. The problem is that sunlight is reflected by the outermost layer of a planet, so one with a thick atmosphere, such as Venus, will yield information only about the composition of its atmosphere.

Even more dramatic was the impact of being able to fly spacecraft near to planets, moons, asteroids and comets. In some cases, they even landed on them (or crashed into them) and sent back information impossible to collect from Earth. Samples have been collected and returned from the Moon, comet Wild2, and asteroid 25143 Itokawa. The quality of images returned by spacecraft orbiting planets is good enough to enable comprehensive mapping of some, but not all of the planets. There are still areas that for now remain unseen, the dark sides of planets or their moons, and those never visited. There is plenty left for future planetary cartographers to do.

Top: *This false-colour view of Mercury gives information about the planet's surface composition.*

Right: *Charon, a moon orbiting the distant dwarf planet Pluto, is a cold, dark, rocky world far from the Sun.*

An artist's impression of the start of the solar system, with a young Sun and protoplanetary disk.

BUILDING A SOLAR SYSTEM

One of the aims of mapping the planets is to discover more about the history of the solar system and how it came to be. Our current understanding is that the Sun, like other stars, formed from a collapsing cloud of gas and dust. As the particles were drawn closer together by their gravity, their collapse accelerated until the centre of the cloud was so dense that nuclear fusion began. The central ball rotated, spinning the left-over material into a disk which circled the star's equator. In this protoplanetary disk, the heavier elements and rock dust were drawn towards the centre by the gravitational attraction of the Sun, while the lighter elements ended up further out. Then a similar process of clumping occurred on a smaller scale. The rocky material nearest the Sun formed the four terrestrial planets, Mercury, Venus, Earth and Mars. The gases and ices flung far from the Sun formed the gas and ice giants, Jupiter, Saturn, Uranus and Neptune.

BODIES BIG AND SMALL

The solar system is home to far more than just planets. Many of the planets have moons (natural satellites), which either formed with them, were created in collisions (as ours probably was) or were captured when wandering asteroids strayed too close to a compelling gravitational field. Many moons have prograde orbits, which means they move around their planet in the same direction as the planet's rotation. They are usually close to the planet and formed with it; they are called regular moons. Retrograde moons orbit in the direction opposite to the planet's rotation, often at a considerable distance away from it. All retrograde moons were captured, and are called irregular moons.

SIZE AND DISTANCE

Although we will focus on mapping individual bodies, it is worth pausing for a moment to put the planets in context. Jupiter is more than eleven times the size of Earth, and the Sun is ten times larger than Jupiter. If Earth were 1 mm across, Pluto would be 0.6 km away! For this reason, astronomical maps never show distances to scale (except occasionally to logarithmic scales); two planets drawn to scale wouldn't fit on the same page. Remember, too, that every map of the planets is a moment frozen in time. The planets and other bodies are constantly hurtling through space, so Earth may be near to Mars or on the other side of the Sun from Mars at any particular moment. Planets don't have circular orbits, so even the diameter of their orbit must be given either as an average or a range, with perihelion (the closest point) and aphelion (the greatest distance).

The relative sizes of the planets. From top: Neptune, Uranus, Saturn, Jupiter, Mars, Earth, Venus, Mercury.

ROCKY PLANETS

WORLDS OF STONE AND WATER

The four rocky planets are closest to the Sun and include our own. As our nearest neighbours, Mercury, Venus and Mars were the first focus for exploratory space probes. They have sparked the imaginations of science fiction writers and artists who have dreamed of populating the solar system.

Left to right: *Mercury, Venus, Earth and Mars. All four rocky (terrestrial) planets are dense, with a solid surface of silicate rocks and a core of heavy metal, mostly iron. There the similarities end, for they are very different worlds.*

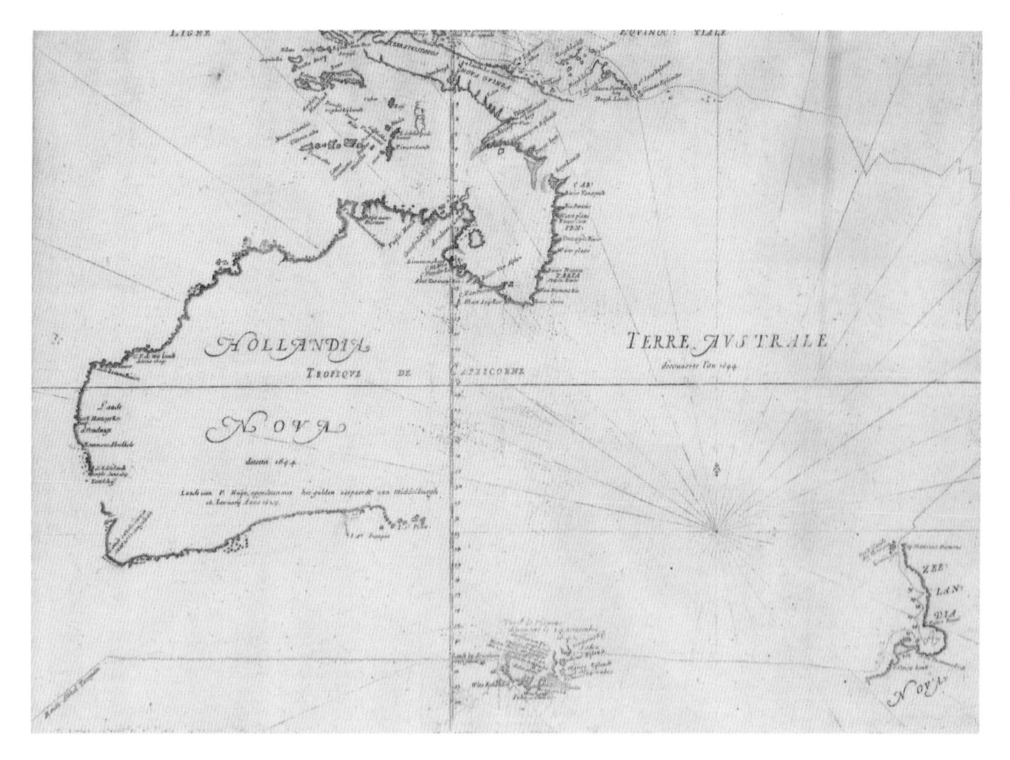

Thevenot's map of Australia made in 1663 acknowledges with large blank areas that the full geography of the area was unknown. More fanciful cartographers sometimes added fantasy beasts to indicate unknown regions.

NEAR AND FAR

The earliest maps of Earth are of a small locality, mapped by people who explored the planet on foot or by simple boat. They had no idea how their region fitted into a larger world. Over time, mapmaking was able to take a wider view. Some early maps position the mapmaker's domain within a wider area of unknown territory; eventually the unknowns were filled in.

Mapping the other bodies of the solar system, rather than their movements, is a recent enterprise which was first made possible by the invention of the telescope. Instead of being constrained to look locally until technology enabled a more remote view, we have been constrained to look from a distance until technology enabled a close-up.

The first maps of the rocky planets were drawn from what astronomers could see through telescopes. The view was of patches of colour, light and shade. There was no way of matching what astronomers saw to features, so understandably there were mistakes: lines on the surface of Mars were mistaken for canals, for instance. (In fact, the lines were not even there – but that's a different point.) The cloudy atmosphere of Venus led people to believe it could be a steamy, tropical paradise. It isn't – its surface conditions make it the hottest planet in the solar system, a hellish cauldron.

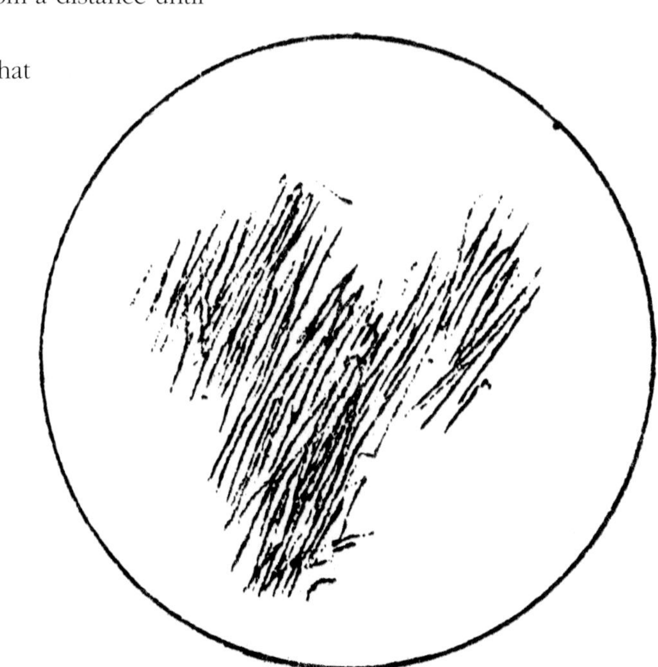

The Dutch astronomer Christiaan Huygens made this sketch of Mars in 1659. It is the first record of any kind of detail on the planet's surface.

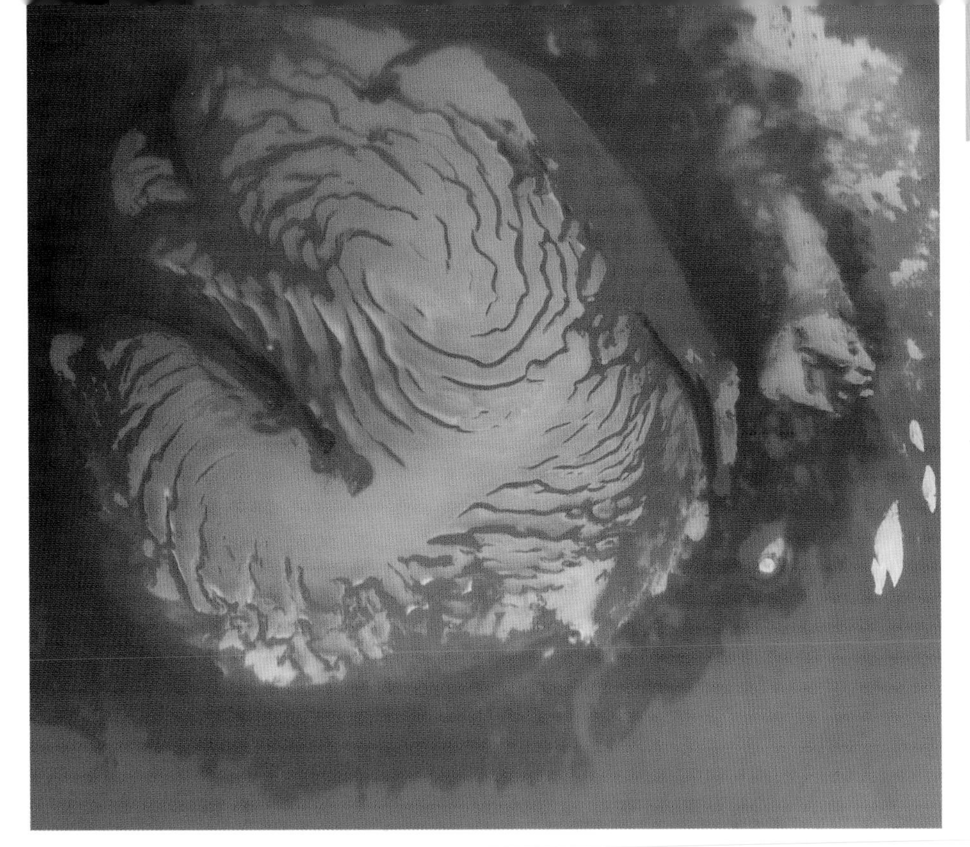

The Hubble Space Telescope clearly shows a polar ice cap on Mars in 1997. Giovanni Cassini first reported ice caps on Mars in 1666.

Improved knowledge of the rocky planets came first with better telescopes, but space-based telescopes, radar and computer imaging have revolutionized planetary mapping. Optical telescopes and cameras used from space have given us stunning pictures of the planets. In the case of the rocky planets, landers have given us close-up views of surface features on Mars, the Moon and, to a lesser degree, Venus. A rover on Mars is the equivalent of a prehistoric mapmaker on Earth, surveying just the land it can cover trudging slowly and recording it in meticulous detail.

We can now see parts of Mars in as much detail as we see the Earth. Rovers such as Curiosity rumble over the planet's surface taking photographs which are relayed back to Earth at the speed of light. This panorama shows the foothills of Mount Sharp in the Gale crater.

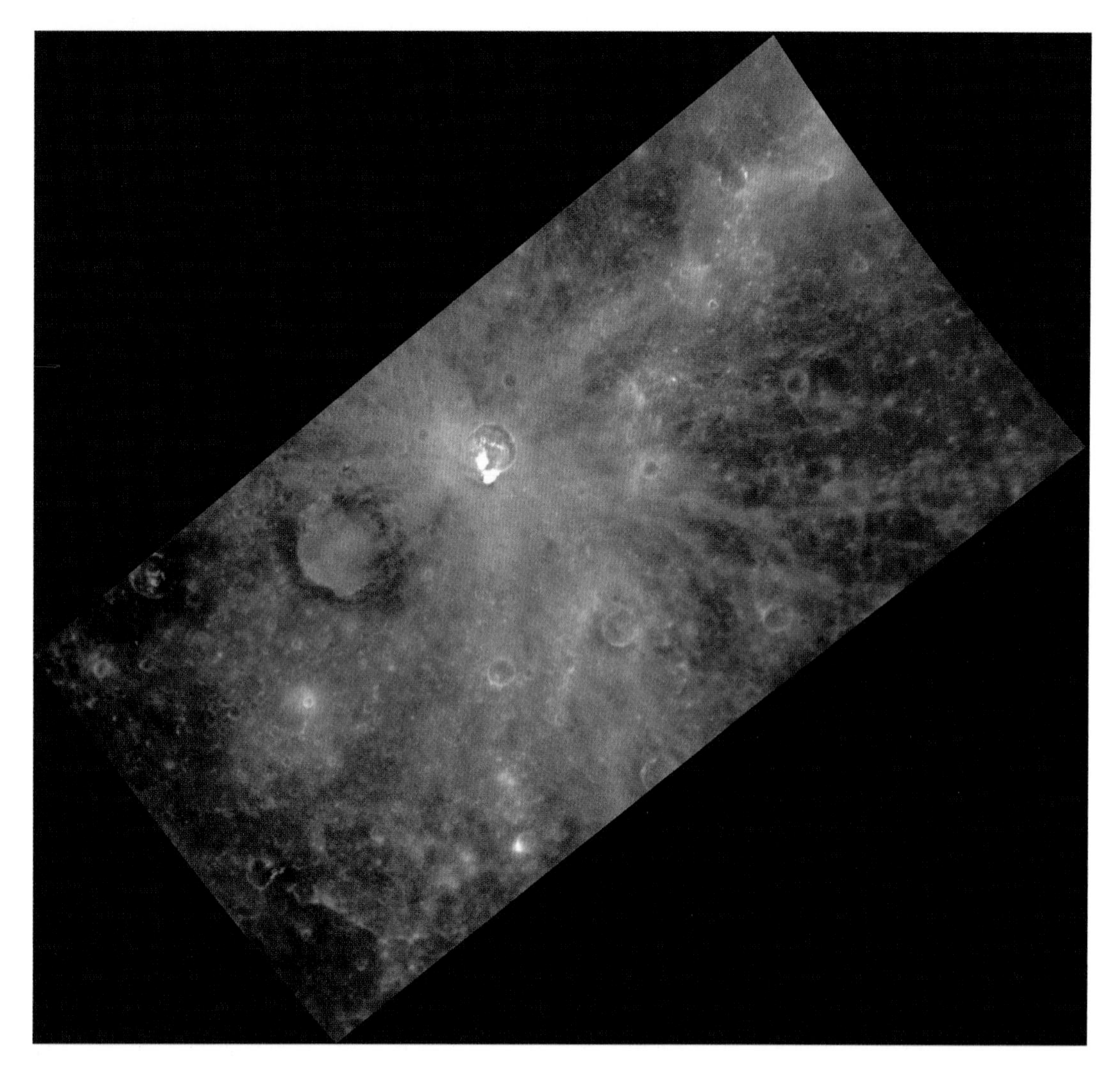

An albedo map of Mercury, with areas of high reflectivity shown in white and areas of low reflectivity appearing dark.

LEARNING FROM LIGHT AND DARK

Improving telescopes first revealed patterns of light and dark on the surfaces of the rocky planets and the Moon. The brightness of light reflected by a planet is known as albedo. On Earth, the places with the highest albedo are the polar ice caps, bright with reflected light, while the sea, darker than the land, has the lowest albedo. On other planets, astronomers use albedo as a guide to geological composition, though it can't alone reveal which chemicals are present. A surface rich in carbon will be dark, for instance, while one rich is silica will be lighter.

EYE IN THE SKY

With space travel, more techniques became available. One of the most valuable for mapping solid bodies is satellite altimetry, which is used to measure the 'lumpiness' of the surface. A craft orbits the planet, bouncing radar off the surface and measuring how long it takes to return. The signal travels at the speed of light, so it's a very brief time. The distance to the surface is calculated from the interval. A computer processes readings from all over the planet, converting the measurements into a map that shows the varying height of the surface – a topographical map, revealing mountains, valleys, canyons and plains.

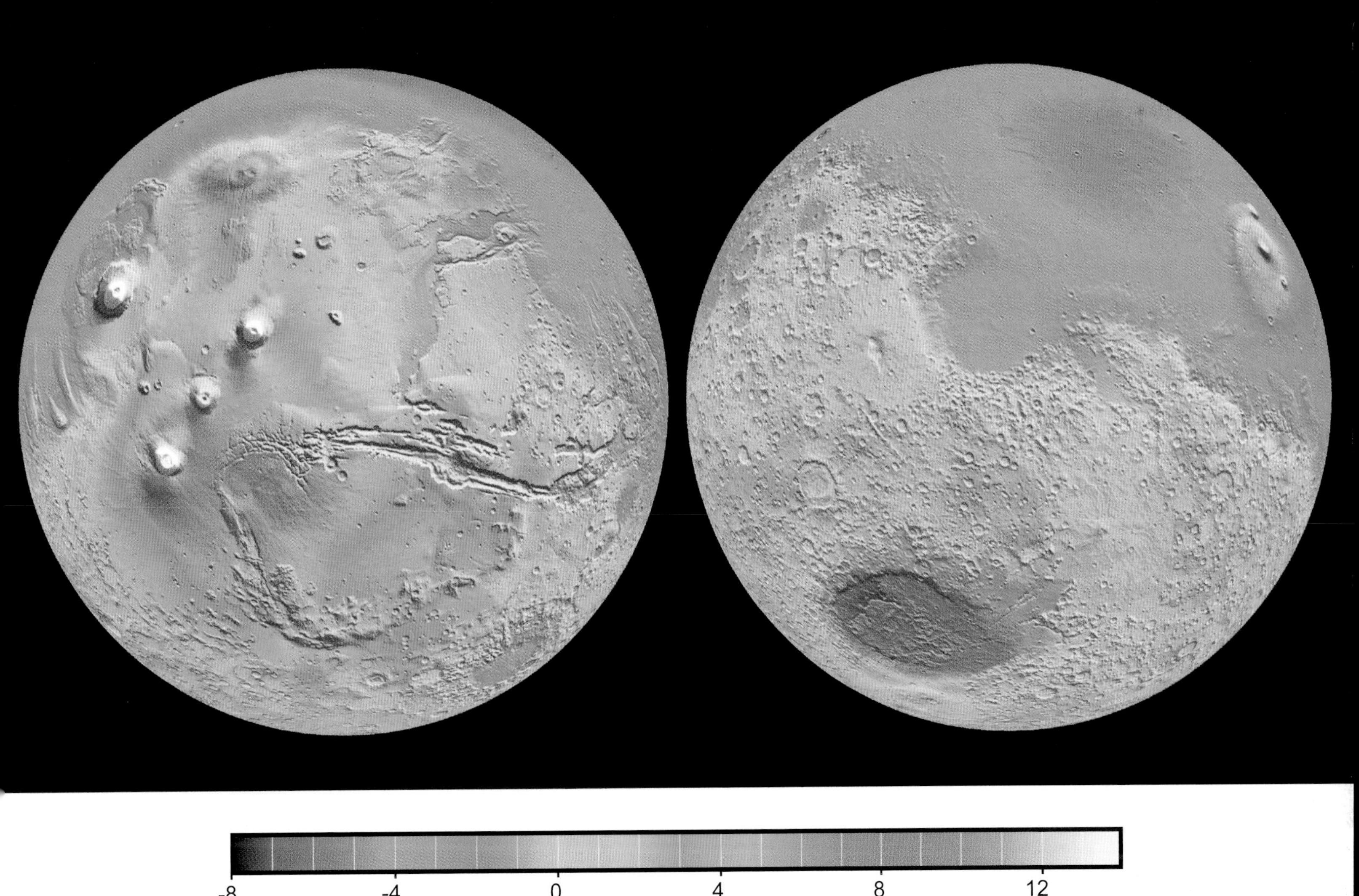

Above: *A topographic map of Mars; the altitude in kilometres of millions of points on the surface was measured by the Mars Global Surveyor Orbiter.*

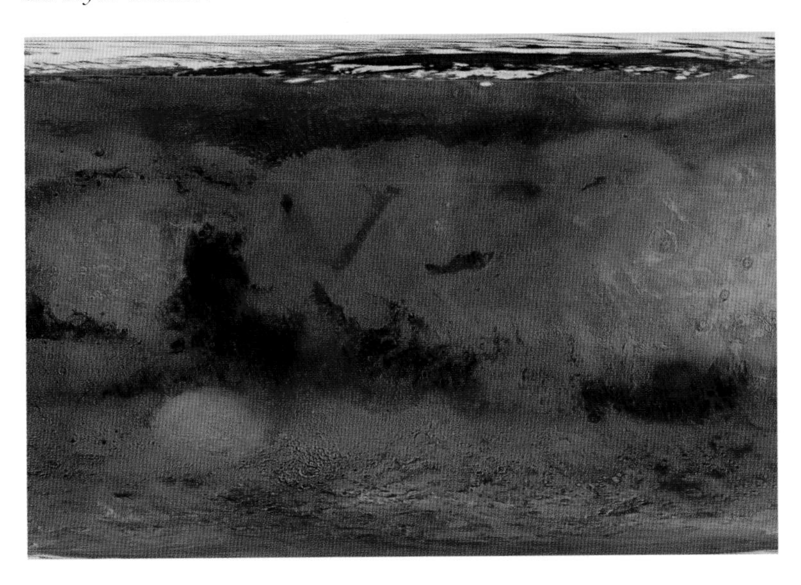

This photo mosaic of Mars is made up of around 100 Viking Orbiter images, taken in 1980.

MIXING PICTURES

Another form of mapping produces photographic mosaics of the surface of a planet. These too are created by spacecraft in orbit. The closer a craft is to the planet's surface, the higher the resolution of the photos it can take (the smaller the area covered by each pixel, so the greater the level of detail).

A photo mosaic, like any other map, can be presented as circular or rectangular, but any flat presentation of a spherical surface involves some kind of distortion. A rectangular map artificially extends the small areas at the poles; circular maps distort areas towards the edges of the circle. Only a three-dimensional globe can accurately represent the surface of a planet.

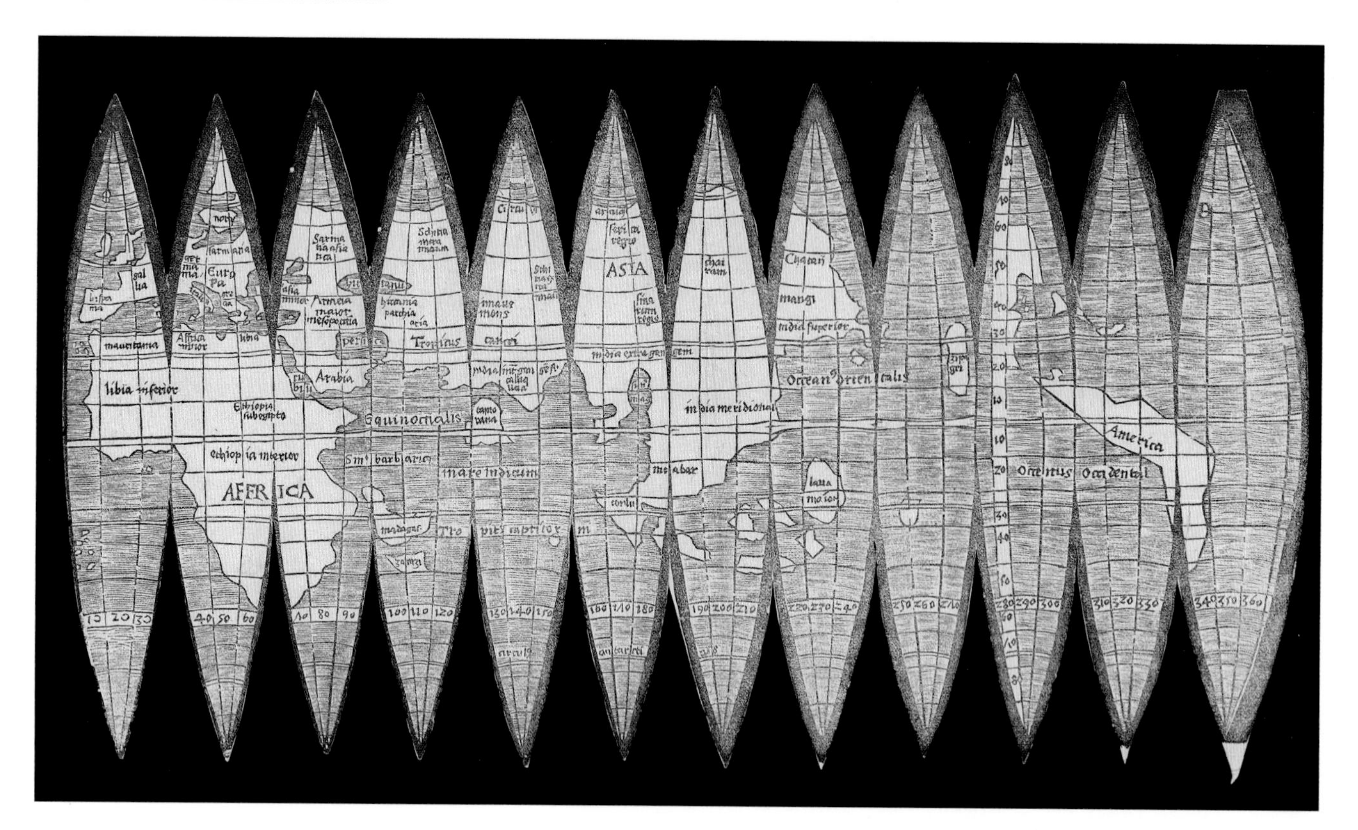

Above: *Martin Waldseemüller produced both a flat world map and these gores for a twelve-part terrestrial globe in 1507. The map was the first to use the name 'America' (visible on the right). The smaller the number of gores, the easier such a map is to read while flat, but the more distortion is involved in representing a territory.*

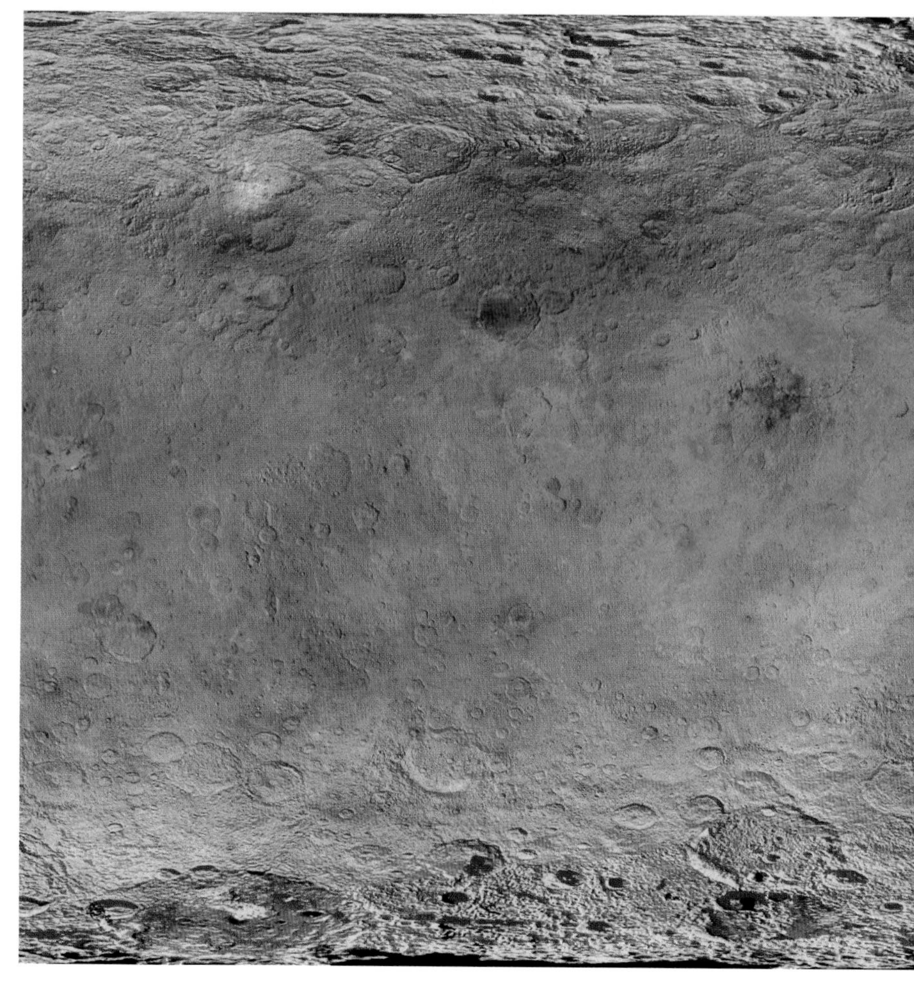

Another method of reproducing the surface of a planet is as a series of 'gores'. A gored map could be cut out and folded around a ball to make a globe, and involves less distortion than a flat map. It is difficult to get an overview of the sphere's surface from this type of map, however.

PLAYING WITH COLOURS

Surfaces are often presented in enhanced or false colour to represent different kinds of information, such as chemical composition or altitude. Sometimes the colour is very clearly not how the planet truly appears; in other cases, it matches our expectations.

A 1992 Magellan image shows, in deep orange, the radio-thermal emissions from an active volcano on Venus.

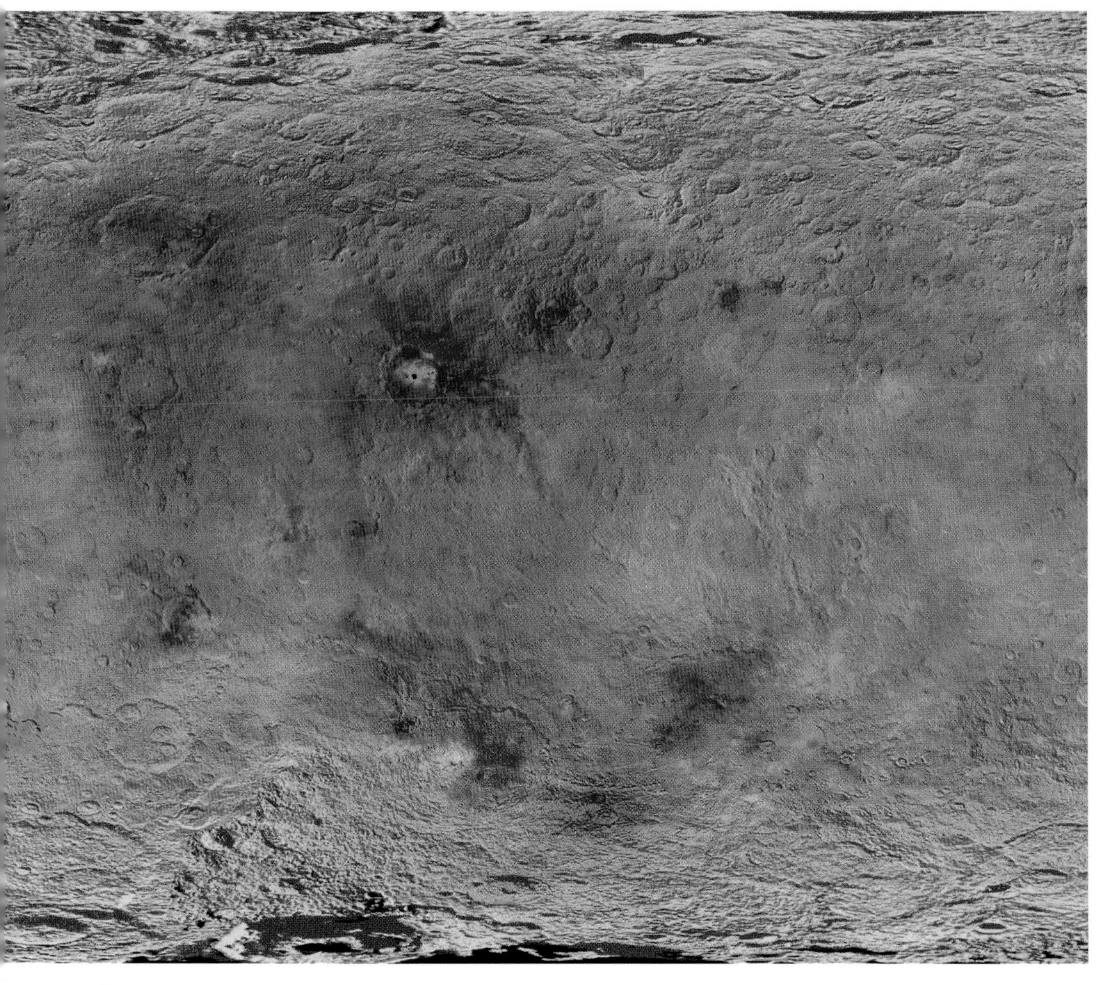

This map-projected view of the dwarf planet Ceres uses false colour to indicate the mineral composition of the surface as well as variations in its age. It is a composite of images taken by NASA's Dawn spacecraft in 2015.

MERCURY

This false-colour image of Mercury shows the mineral composition of the planet's surface.

The surface of Earth retains few scars from its early life, but the surface of a planet such as Mercury flaunts its history. Earth's rocks have been weathered and changed by tectonic activity, obliterating evidence of the many asteroid strikes that the planet suffered billions of years ago. But elsewhere in the solar system this evidence is still plain to see and helps us piece together the history of the planets.

Mercury has many impact craters, produced by rocky bodies crashing into the surface of the planet. Some of these have been filled or partially filled by lava from volcanic activity. Sometimes the solidified lava has then been cratered again by further impacts.

With only a very thin atmosphere, and turning on its axis so slowly that its day lasts 59 Earth days, Mercury experiences extremes of temperature. In full sun, without protection, the surface is scalding hot but, with no atmosphere to hold heat, it soon cools when it is in shade.

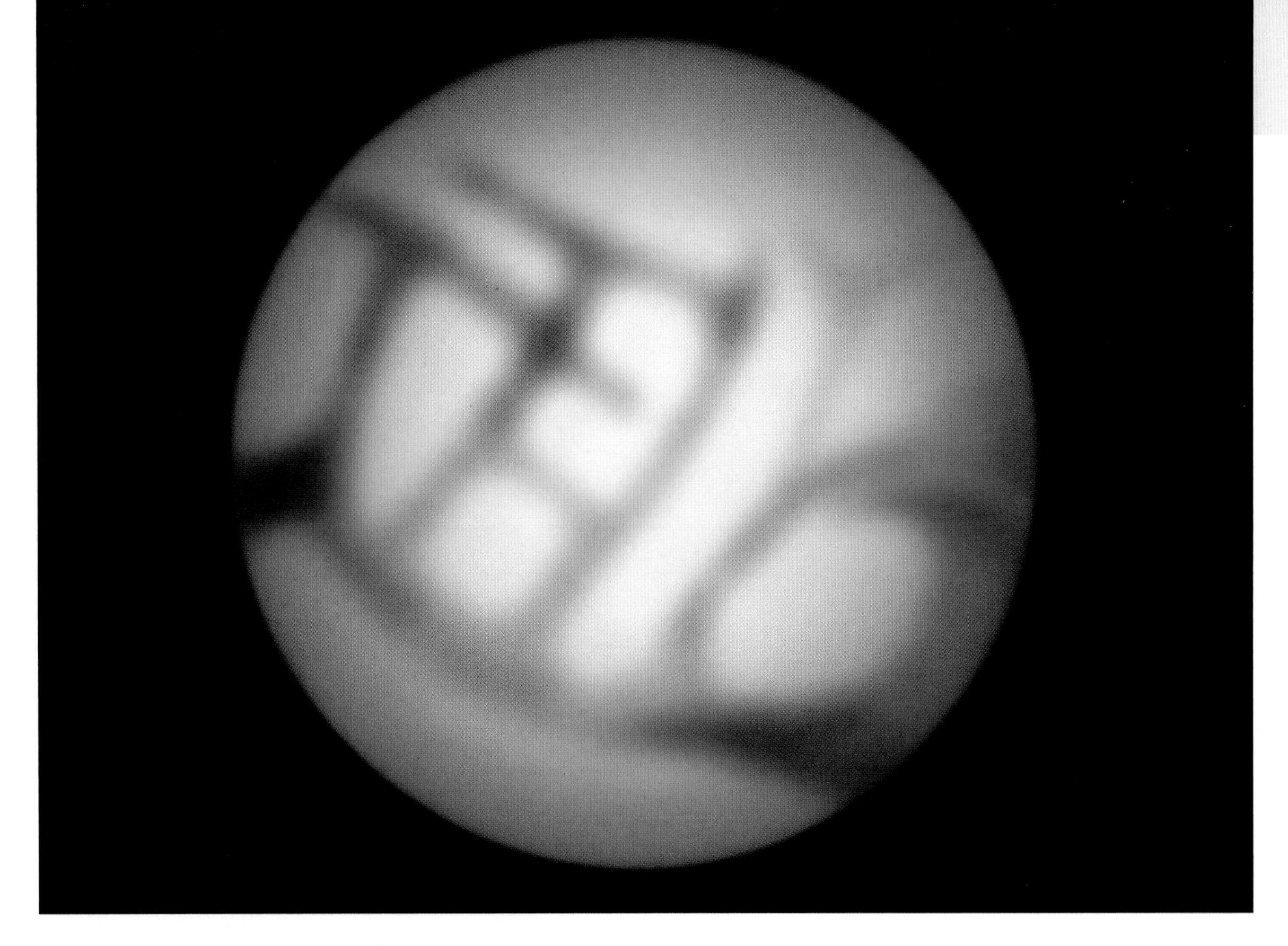

Planetary focus	Mercury
Length of year (orbit around Sun)	88 days
Length of day (rotation on axis)	59 days
Size x Earth mass	0.06
Size x Earth radius	0.4
Average distance from Sun	58 million km (36 million miles)
Moons	0
Discovered	Prehistory
Missions	Mariner 10, NASA, 1974; Magellan, NASA, 2008; Beppi-Colombo, ESA/ JAXA (launched 2018, due to arrive 2025)

ILLUSORY DETAILS

Mercury is always near the Sun, so only visible soon after sunset or before dawn at particular times of the year. This makes it difficult to observe. Although the Italian astronomer Giovanni Zupi discovered in 1639 that Mercury has phases, like the Moon, he wasn't able to make out any features on its surface. No one could, in fact, until another Italian, Giovanni Schiaparelli, produced drawings that show lines on the surface of the planet in 1889. These lines only very roughly correspond with surface features now seen on Mercury. Schiaparelli was good at seeing non-existent lines on planets, most famously on Mars (see page 76). He did, however, devise a coordinate system for Mercury.

The American astronomer Percival Lowell was an enthusiastic hunter of Schiaparelli's planetary lines. He mapped the lines he was sure he could see and concluded that they were rather like wrinkles, caused by the planet cooling and its surface then being overlarge. The surface of Mercury does have ridges which could be the result of such cooling – but Lowell could not have seen them with his telescope.

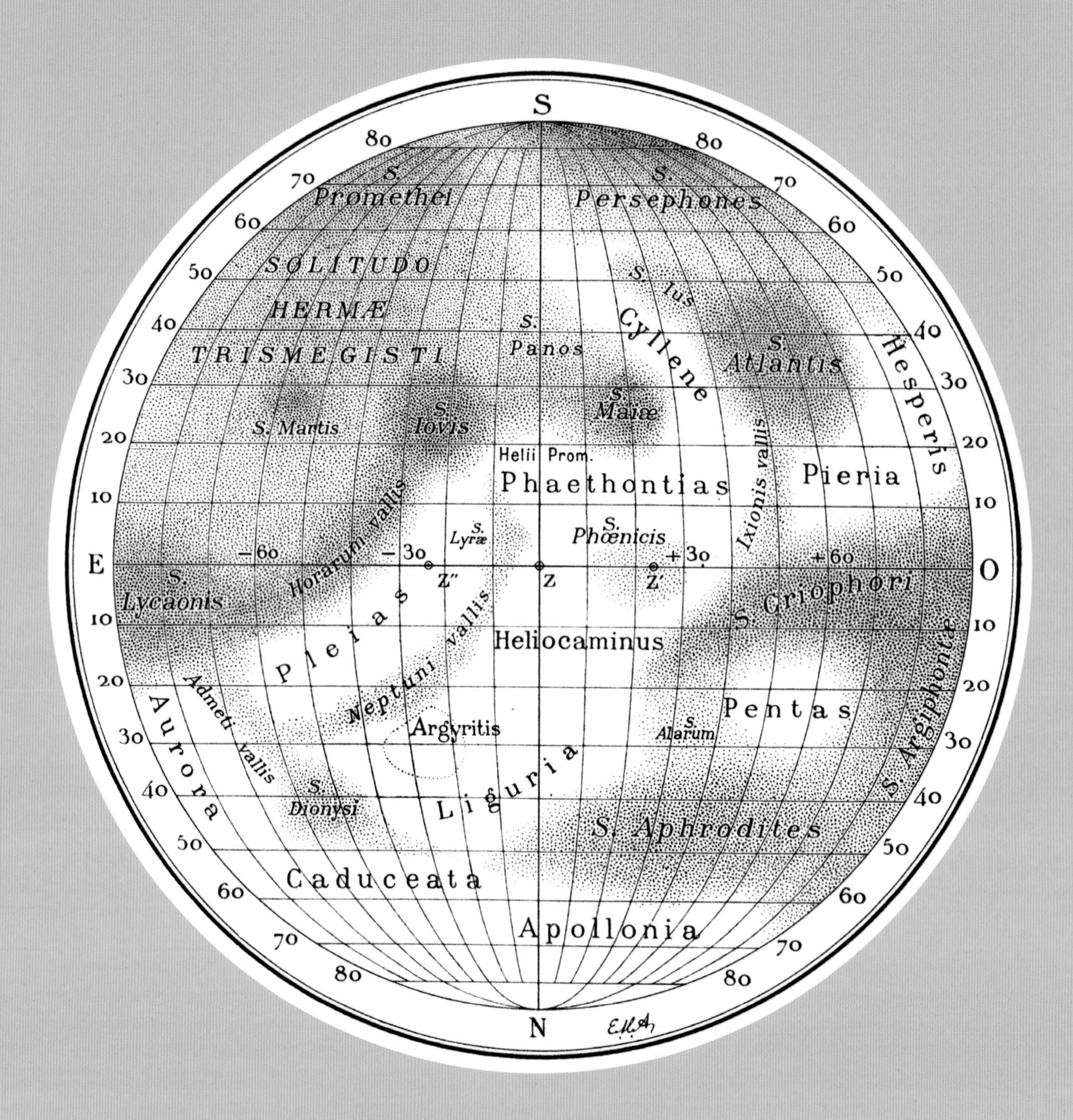

EUGENIOS ANTONIADI, MAP OF MERCURY, 1934

The first map of surface features of Mercury was made in 1934 by Eugenios Mihail (later Eugène-Michel) Antoniadi, a Greek astronomer who spent much of his life in France. He devised a naming convention for the features he saw on Mercury, and this formed the basis of the scheme adopted by the Task Group for Mercury Nomenclature set up in 1973. Antoniadi's observations were not entirely accurate, though, as he wrongly assumed that Mercury is tidally locked (meaning it always has the same side facing the Sun). As Mercury takes 59 days to make a full rotation, observations on consecutive days show a very similar view, but in fact all areas of the surface get their turn in the sun.

NASA, MARINER 10, 1974

Mariner 10 was launched on 3 November 1973 with the aim of investigating the surface, atmosphere and physical characteristics of Mercury and Venus. It was the first craft launched by NASA to fly past the inner planets.

Mariner made three passes by Mercury about six months apart, photographing the surface each time. As the same side of Mercury was visible on each occasion, it was only able to photograph 45 per cent of the surface. Mariner also measured the planet's magnetic field, returning the surprising result that it is very similar to that of Earth, despite Mercury's much slower rotation.

This image is a mosaic made of 18 photos taken by Mariner 10 at an altitude of 200,000 km (124,274 miles). Mercury's north pole is at the top of the picture.

Mariner 10, launched in 1973 to fly by Mercury and Venus.

MESSENGER, GEOLOGY OF MERCURY, 2008–9

The second and so far the only other mission to Mercury was NASA's MESSENGER (MErcury Surface, Space ENvironment, GEochemistry and Ranging), which carried out three series of observations in 2008 and 2009. MESSENGER's mission was to investigate the chemical composition of the surface, the geological history of the planet, the size and state of its core, its exosphere (thin envelope of gases) and magnetosphere (magnetic field). It was the first craft to orbit Mercury (see image on the right) and took enough photographs to map 95 per cent of the surface. These were at a higher resolution than the Mariner photographs, increasing the precision of the mapping.

The false-colour image above indicates the chemical, mineralogical and physical differences between the rocks that make up the planet's surface.

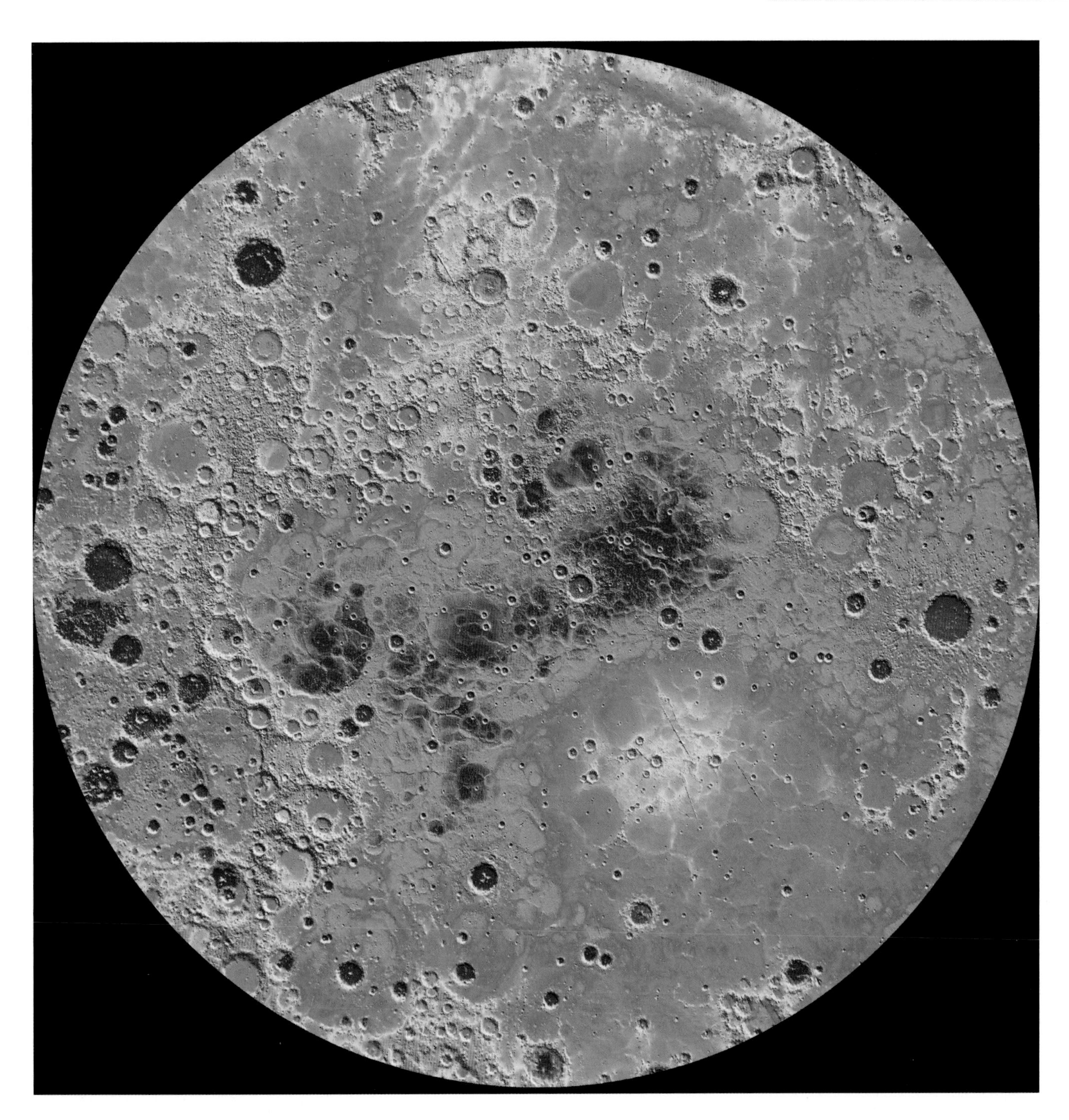

This topographical map of Mercury shows the highest regions in red and the lowest in purple, with a difference between them of approximately 10 km (6.2 miles). The north pole is in the centre; the deep craters near the pole may contain ice.

MESSENGER, MERCURY TOPOGRAPHY, 2008–15

MESSENGER's original photography mission was completed in 2009; extended missions afterwards continued to return data until the craft was deliberately crashed into the planet's surface in 2015. The data and raw photographs will continue to yield new insights for years. Some of the processed images use colour to show the composition, elevation, presence of water and other features of the planet.

MESSENGER, MERCURY GRAVITY FIELD, 2015

This image shows gravitational anomalies which reveal information about the structure of Mercury beneath its surface. The red areas denote concentrations of gravity produced by denser subsurface material. The large central area is the Caloris basin, Mercury's largest crater; the red region on the right is the Sobkou area. The north pole is towards the top.

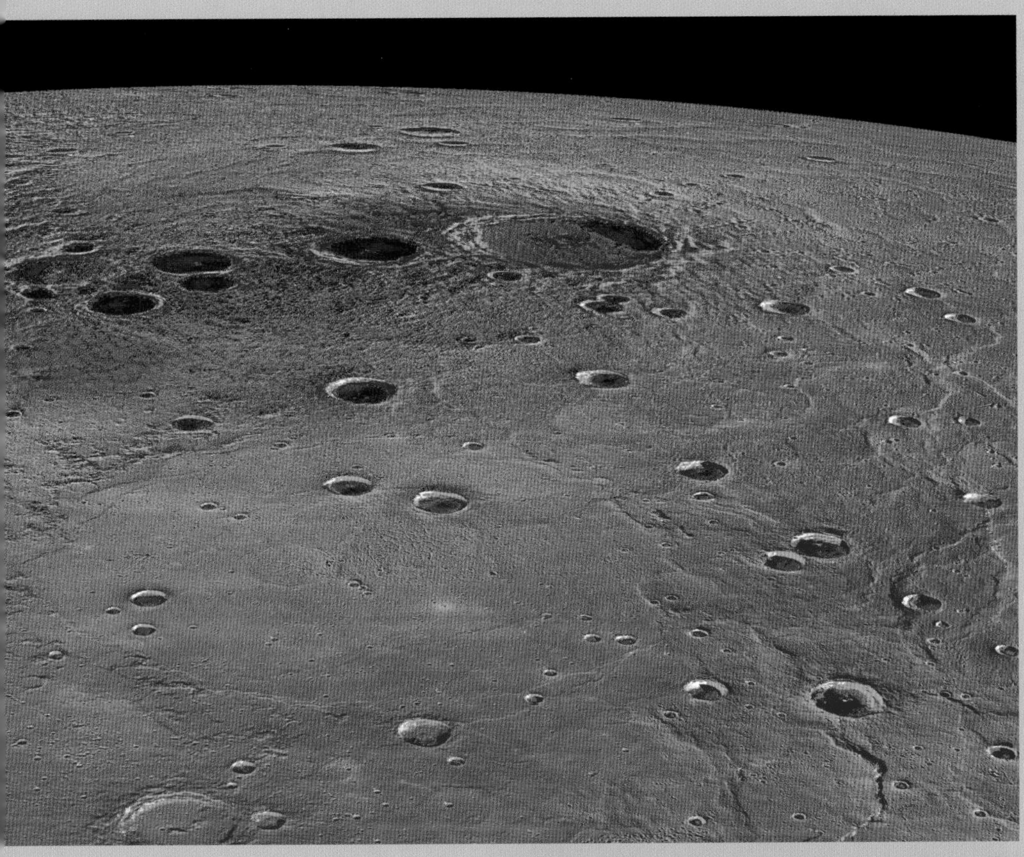

MESSENGER, MERCURY'S NORTH POLE, 2015

This map of Mercury's north pole was made in 2015. It is coloured to show the temperature gradient, with red hottest and purple coldest. The area was in full sun at the time. The temperature in the red areas is well over 100° C. Some parts of the craters are in permanent shadow; the temperature here can be as low as -220° C. Some craters harbour water ice that has probably been frozen for billions of years.

MESSENGER, MERCURY'S CLIFFS

Images from MESSENGER reveal cliff-like scarps on the surface of Mercury which are thought to be the result of the planet shrinking as it cools. As the interior shrinks, the surface has to wrinkle to fit the smaller subsurface.

VENUS

With its thick atmosphere of greenhouse gases, Venus is much hotter than Mercury though further from the Sun. It was the first planet visited by spacecraft, but little could be seen of it in earlier centuries. Like Mercury, it is visible only at dawn or dusk.

The first successful mission to Venus was NASA's Mariner 2 fly-by in 1962. It sent back startling details of a hot planet under a crushingly dense atmosphere. Before this, some people had hoped Venus would be like Earth, perhaps with better weather as it was closer to the Sun. But, no – it has a surface temperature hot enough to melt lead, at 467° C (872° F), an atmosphere that is 95 per cent carbon dioxide, with clouds of sulphuric acid and an atmospheric pressure 75–100 times that on Earth. Venus is one of only two planets that rotate the 'wrong' way (east to west); the other is Uranus. At an early point in its history, Venus might have been knocked over, effectively rotating through 180 degrees north–south so that its north pole is now at the bottom. It has little axial tilt (only three degrees), so doesn't have distinct seasons. Like Earth, it has a metallic core, a rocky mantle of magma and a solid crust.

SOLAR AND SIDEREAL DAYS

The solar day is the time it takes a planet to rotate on its axis so that Sun appears in the same place in the sky. The sidereal day is the time it takes a planet to rotate so that its position is the same relative to the fixed stars. Because Venus rotates in the opposite direction to its orbit around the Sun, and rotates slowly, its sidereal day is much longer than its solar day.

Planetary focus	Venus
Length of year	225 days
Length of solar day	117 days
Length of sidereal day (rotation on axis)	243 days
Size x Earth mass	0.8
Size x Earth radius	0.9
Average distance from Sun	0.7 AU; 108 million km (67 million miles)
Discovered	Prehistory

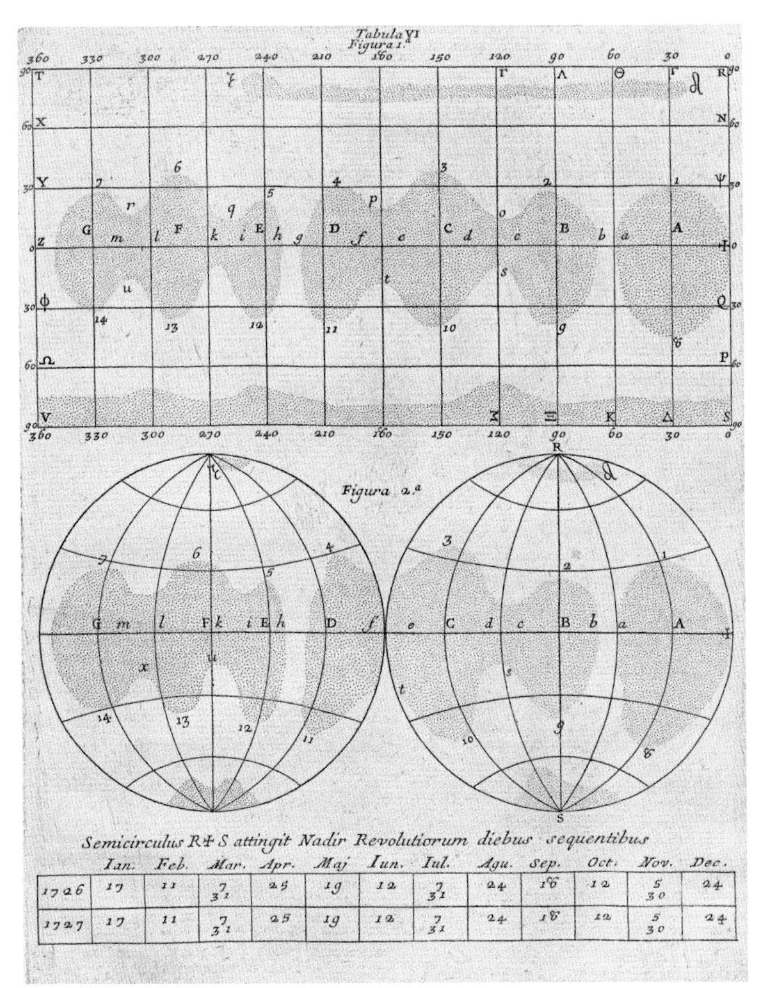

FRANCESCO BIANCHINI, VENUS MAP, 1728

Because Venus is shrouded in thick cloud, it is impossible to make out any surface features with an optical telescope. But early astronomers were certain they could see features. The Italian scientist Francesco Bianchini identified light and dark areas on the planet which he fancifully interpreted as oceans and continental landmasses. Other astronomers claimed to see mountains on Venus, and Francesco Fontana was certain he had observed a moon. Bianchini's areas of light and shade, if they existed, will have been banks of thicker and lighter cloud.

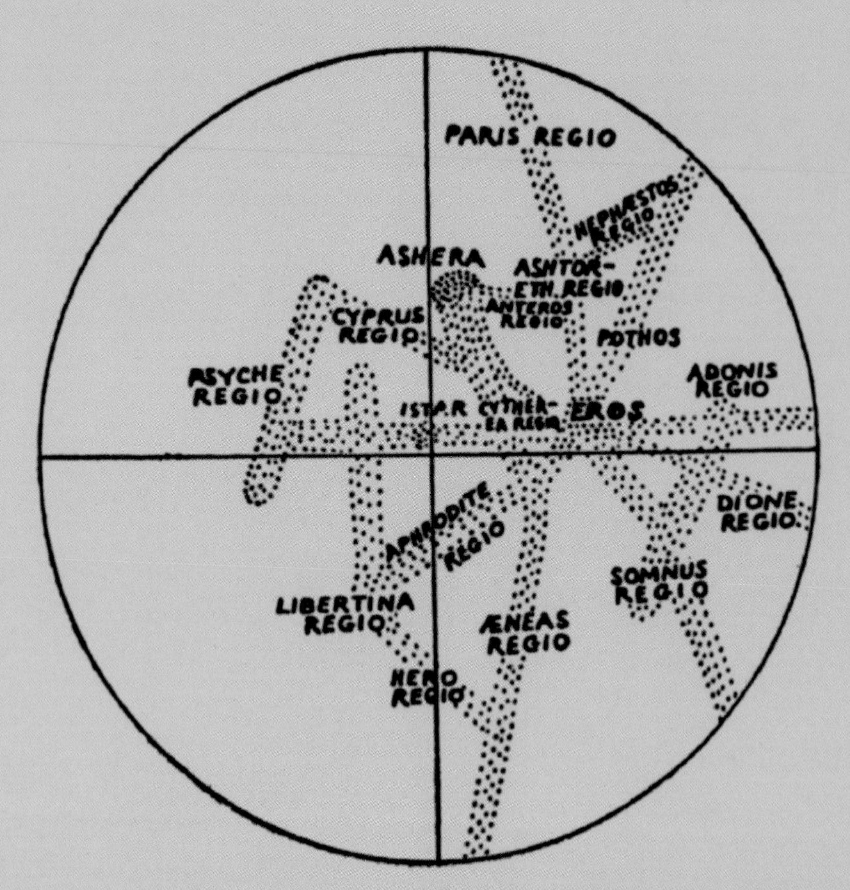

PERCIVAL LOWELL, LINES ON VENUS, 1896

More than 150 years after Bianchini, Percival Lowell saw lines on Venus as he had on Mercury. No other astronomer had ever seen anything similar. Lowell briefly retracted his claims in 1902, but returned to them the following year.

It turns out that Lowell's map of Venus is a map of the inside of his own eyeball. As Venus is so close to the Sun, Lowell had reduced the aperture on his 24-inch telescope to just 3 inches, or perhaps even less. The result was that he was effectively shining a bright light into his eye. Lowell saw the shadows of the blood vessels in his retina superimposed on the image of the planet.

APPROACHING VENUS
The first photograph of Venus (not shown here) was taken by Mariner 10 in February 1974 as it approached the planet. It shows thick banks of swirling clouds that obscure the surface. The ultraviolet image is colour-enhanced to show the atmospheric effects more clearly. The clouds of Venus extend 70 km (43.5 miles) above the surface. A giant streak discovered in the clouds in 2019 suggests that jet streams similar to those on Earth are present, but mapping the weather of Venus is still at an early stage.

VENERA 9 AND 10, SURFACE OF VENUS, 1976

A series of Soviet Venera missions dropped probes into Venus' atmosphere. Some landed successfully on the surface and returned data before falling prey to the hostile conditions. The images returned by the Venera craft are the first photographs from the surface of another planet. Venus is difficult to explore because of its hostile conditions. Rovers can amble around Mars for years taking photos, digging out samples and recording data, but no probe has survived longer than two hours on Venus. The probe has to collect and transmit all its data before being destroyed by the hot, acidic atmosphere and crushing pressure. There has been no return to the surface of Venus since the 1980s.

VENERA 13, FIRST COLOUR PHOTO OF THE SURFACE OF VENUS, 1982

The sky on Venus looks yellowish because of the sulphur in the atmosphere. The rocks are grey and black, but light filtering through the clouds gives the planet a yellow/orange tinge as seen in the first colour photographs from the surface. Parts of the spacecraft are visible at the bottom of the image.

MAGELLAN, RADAR-DERIVED MAP OF VENUS, 1990

The NASA Magellan mission orbited Venus in 1990. Its main purpose was to use radar to derive a topographic relief map of the surface. The thick blanket of clouds made optical imaging of the surface impossible, but radar can cut through the clouds. It took four years to complete the mission; each imaging cycle took 243 days – the time it took for Venus to complete one revolution beneath Magellan. The result was a map covering 98 per cent of the planet's surface to a resolution of 30 km (18.5 miles).

The different features of Venus – mountains, plains and valleys – are clearly visible. Bright areas are high altitude. The surface is thought to be only around 500 million years old and around 85 per cent of it is characterized by volcanic lava flows. The high surface temperature makes it likely that lava can travel much further than on Earth before solidifying. Although Venus has an atmosphere, with clouds and wind, the lack of water means the surface weathers only slowly. The bright region is Maxwell Montes; the north pole is at the centre of the image.

SURFACE DETAILS FROM MAGELLAN DATA
Topographic data from the Magellan mission has been used to produce detailed images of surface structures on Venus. Although these are obscured by the clouds swirling around the planet, we can create images from radar data that contain the same information as photographs.

MAGELLAN, THE MARGARET MEAD CRATER, 1990

There are only about 1,000 known impact craters on Venus. Margaret Mead is one of the largest, at 275 km (170 miles) in diameter. It's a multi-ring crater with the innermost scarp probably representing the original cavity. The lighter crater bottom suggests it was flooded either with melted rock from the impact or with lava from volcanic activity. There are no small impact craters on Venus comparable to those on the Moon and Mercury. The very thick atmosphere would prevent any but the largest meteors reaching the surface; smaller ones would burn up on the way down.

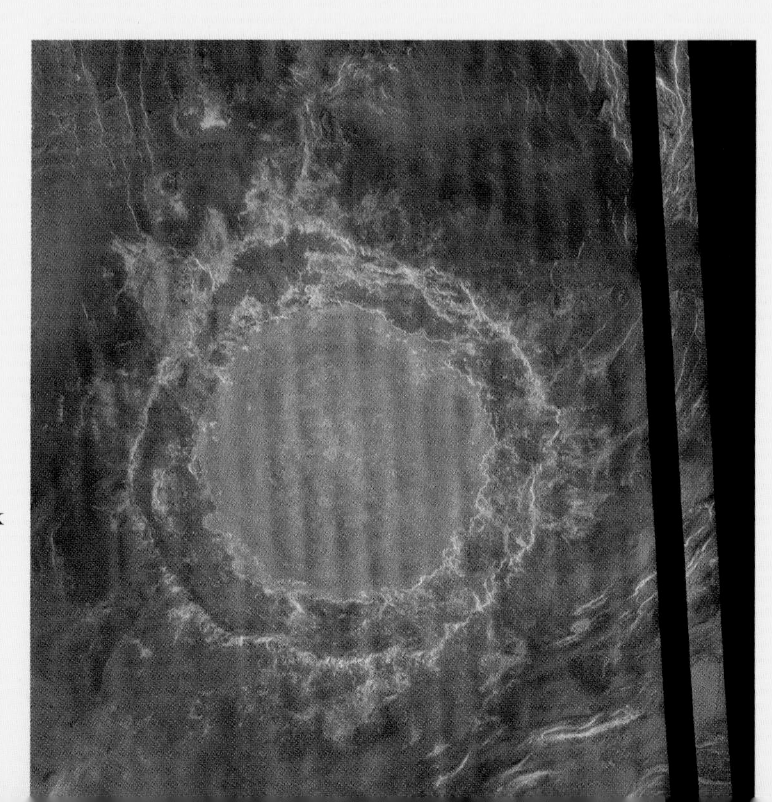

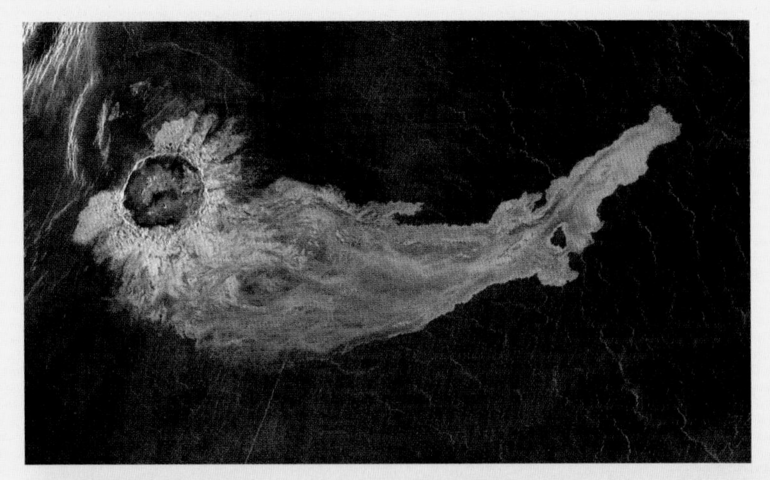

MAGELLAN, THE ADDAMS CRATER, 1990

In the Aino Planitia region of Venus, the Addams crater demonstrates clearly how the planet's high surface temperature allows lava flows to continue unimpeded for long distances. The crater is 90 km (56 miles) across and the lava flow extends for 600 km (373 miles). Magellan found thousands of volcanoes and volcanic plains on Venus.

NASA, MAAT MONS, 2004

Combining topographic data from Magellan and colour data from the early Venera photos, computer imaging can create three-dimensional views of features as they would appear if you could stand on the surface of Venus.

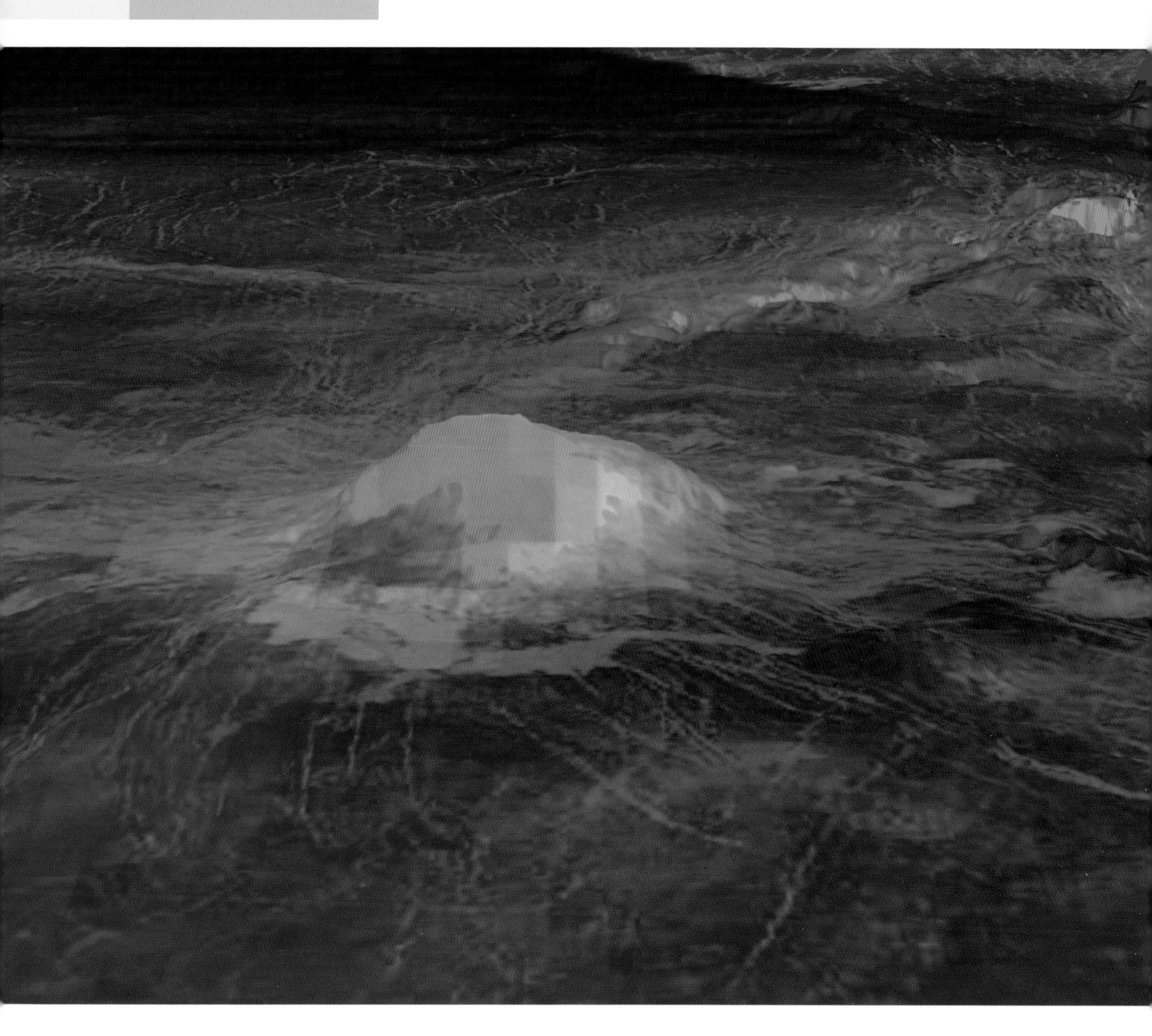

ESA, THERMAL LANDSCAPE OF IDUNN MONS, 2010

Using computers to combine data of different types produces a rich map of *Idunn Mons* in the *Imdr Regio* of Venus. Radar data from Magellan shows areas of rough (dark) and smooth (light) ground and demonstrates altitude. The coloured overlay is thermal data from Venus Express. The temperature shows that *Idunn Mons* is volcanic. Neither set of data alone would show that this is a volcanic mountain: it's impossible to tell just from the thermal data whether hot-spots are mountains or rifts, or to tell from the topographic data alone whether the mountains are volcanic.

MAGELLAN, TOPOGRAPHY OF VENUS, 1994

These maps of both hemispheres of Venus were based on the Magellan topographic data. Gaps have been filled by radar data from the Arecibo telescope on Earth and with information from Pioneer and Venera spacecraft. The image on the right is centred on the south pole.

The image below shows a huge mountain chain. While low-altitude terrain on Venus is mostly dark basaltic rock, the highlands are frosted with snow. Metals vaporize on the hotter plains and condense at altitude, falling as metallic snow on the mountains.

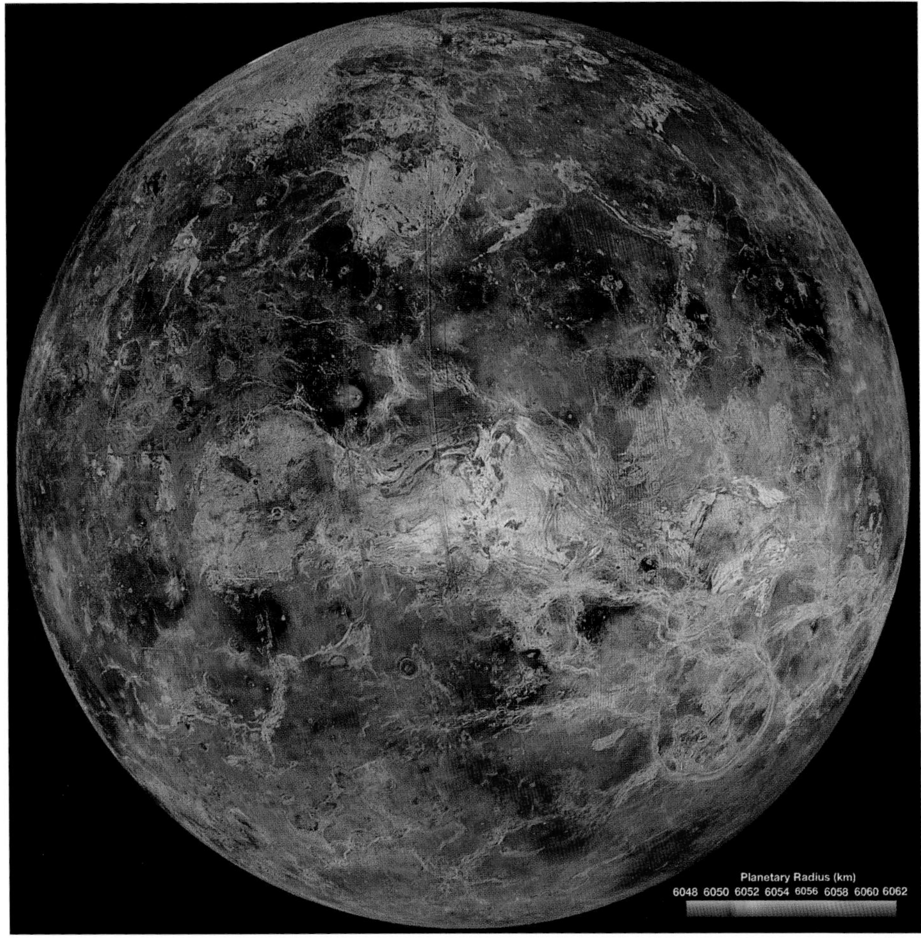

VENUS EXPRESS, CLOUDS AROUND VENUS, 2006

The ESA Venus Express mission arrived at Venus in 2006 and studied the planet until 2014. It used imaging equipment that worked in visible and infrared light, and spectroscopic equipment to determine the composition of the clouds. It found a sudden spike in the amount of sulphur dioxide in the Venusian atmosphere in 2006. On Earth, sulphur dioxide is produced only by volcanic eruptions, so this suggests current or very recent volcanic activity on Venus.

Mapping the movement of clouds around Venus' south pole using infrared reveals a vortex (below). Clouds move very quickly above Venus, circling the planet in about four Earth days.

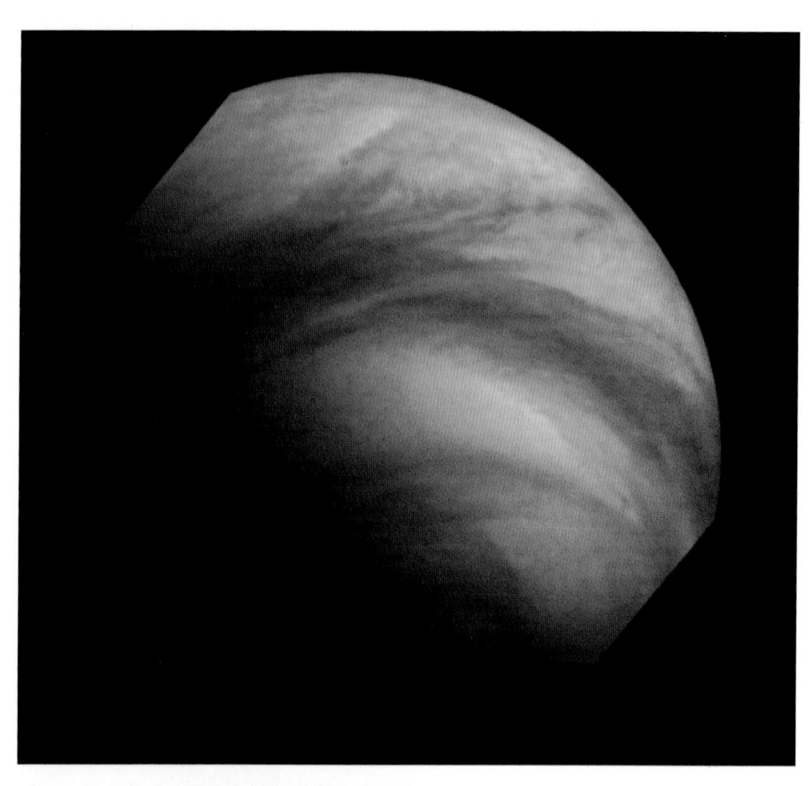

AKATSUKI, VENUS, 2016

The Japanese mission Akatsuki is studying the atmosphere of Venus, in particular the dynamics which mean that while the planet spins at 6 km (3.7 miles) per hour at its equator, the atmosphere spins at a sprightly 300 km (186 miles) per hour. These false-colour images show the night and day sides (right and below respectively) of Venus. The darker areas in the night side photograph represent thicker cloud. The image of the day side reveals differences in the components of the Venusian atmosphere. Most of the cloud is sulphuric acid, produced from sulphur dioxide rising through the atmosphere, and an unknown chemical that is absorbing ultraviolet. Both images show that the equatorial region is turbulent and stormy while the polar regions are calmer.

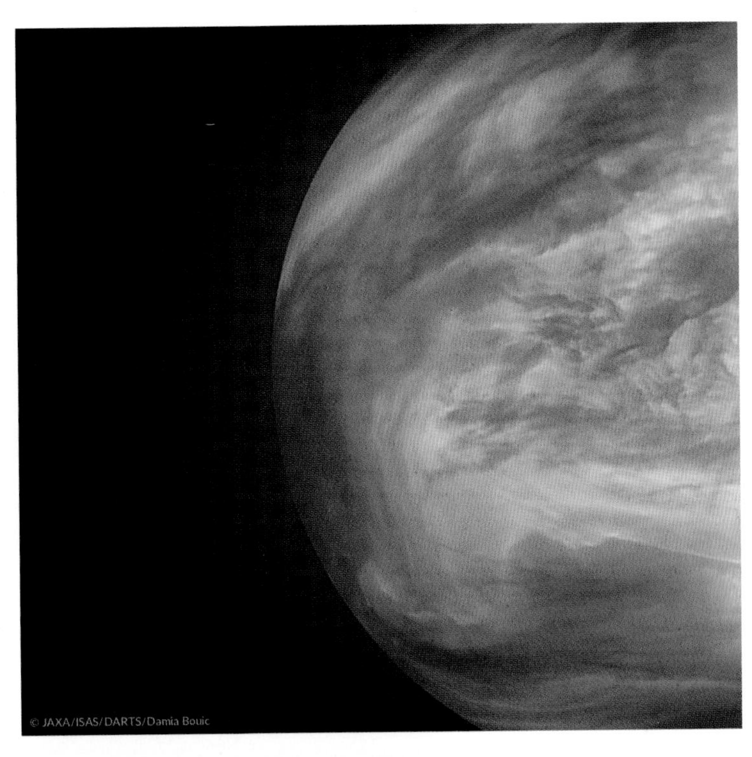

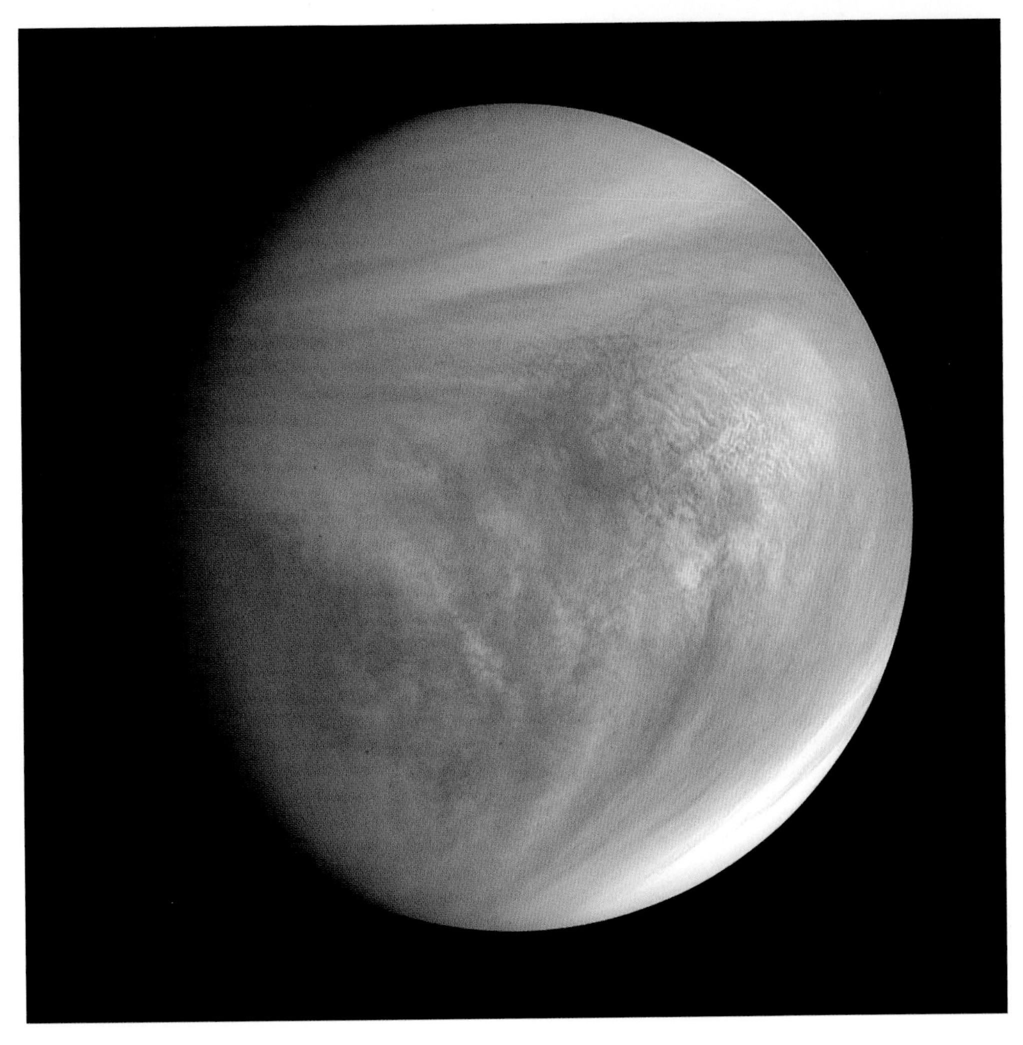

EARTH

Even if Earth were not of particular interest to us because we live here, it would undoubtedly be the most interesting of the rocky planets. With its mix of solid land and oceans of water, polar ice caps, an atmosphere with clouds and, of course, bountiful life, it would be the most rewarding for any visitor to the solar system to study.

Earth photographed from space; the landmasses of North and Central America can be seen beneath the swirling clouds.

Planetary focus	Earth
Length of year	365.25 days
Length of day	24 hours
Average distance from Sun	150 million km (93 million miles)

Earth is naturally the most intensively studied and extensively mapped of the planets. The most inaccessible places can be mapped from the sky, and from space, just as we map other planets. The beds of the oceans are mapped with sonar, working in the same way as the radar mapping of other worlds. But in addition to mapping Earth's geological features, gravity, temperature and composition, we can map human settlements and the distribution of plant and animal life. We can also compare our current world with other maps, and even some retrospectively drawn maps, to see how it has changed.

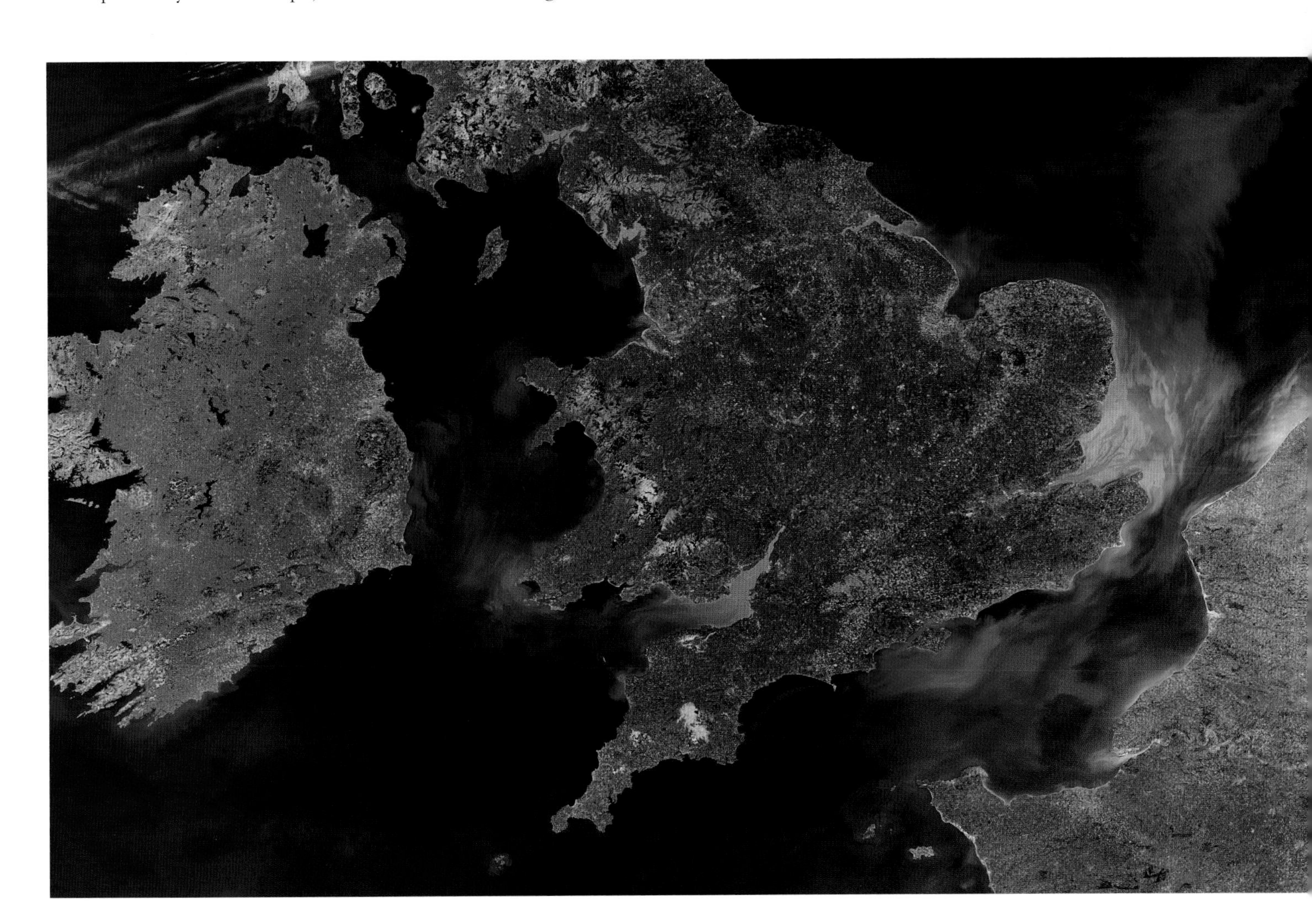

Below: *The earliest known world map was made on a clay tablet in Babylon (ancient Iraq) in the 6th century* BC. *The River Euphrates runs from top to bottom (north to south), its mouth labelled 'swamp'. Named areas and cities are Assyria, Susa, Urartu, Habban, Bit-Yâkin (the original area occupied by the Chaldeans) and Babylon. The known area is surrounded by a circle labelled 'bitter river', assumed to be the ocean.*

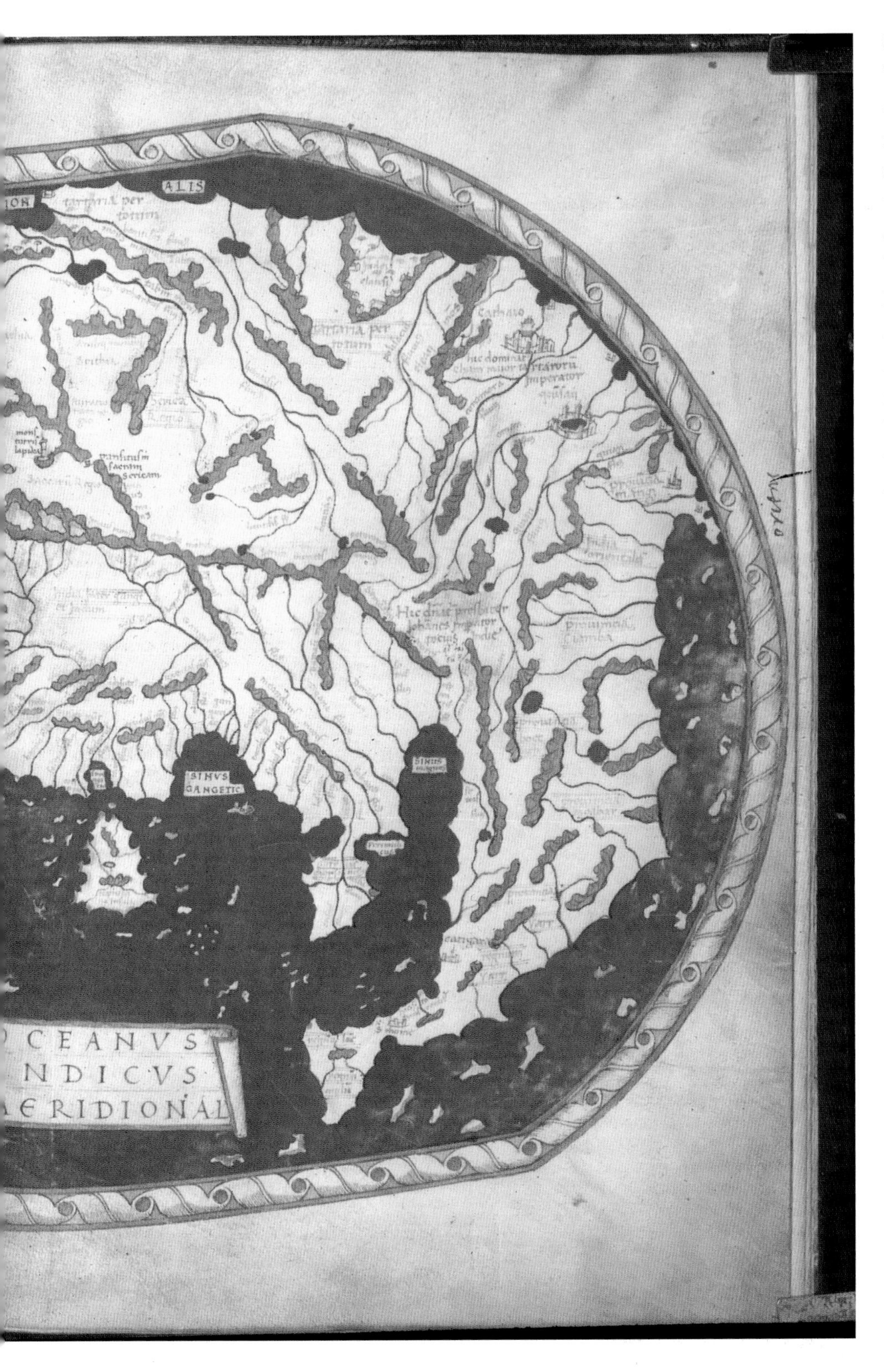

STARTING CLOSE TO HOME

The very earliest maps of Earth show small, local areas. Prehistoric mapmakers had no way of getting an overview of their locality within the larger picture of the landmass on which it was situated. The best they could do was to go to high ground to see how their settlement and landmarks fitted into a landscape of rivers, hills, mountains and forests. Those with boats could trace out the line of the coast, though it's quite hard to draw a map that way and not surprisingly the coastline shown on early maps is a poor match to the real coastline.

Early maps often showed awareness of the limits of the cartographer's knowledge. There are shady areas beyond the margins where the terrain is unknown, often labelled 'terra incognita'. Sometimes these lands and seas are dotted with imaginary monsters. At other times, mapmakers assumed they knew the full extent of the world.

The Martellus map of 1490 shows the whole world as it was known – without the Americas or Australasia.

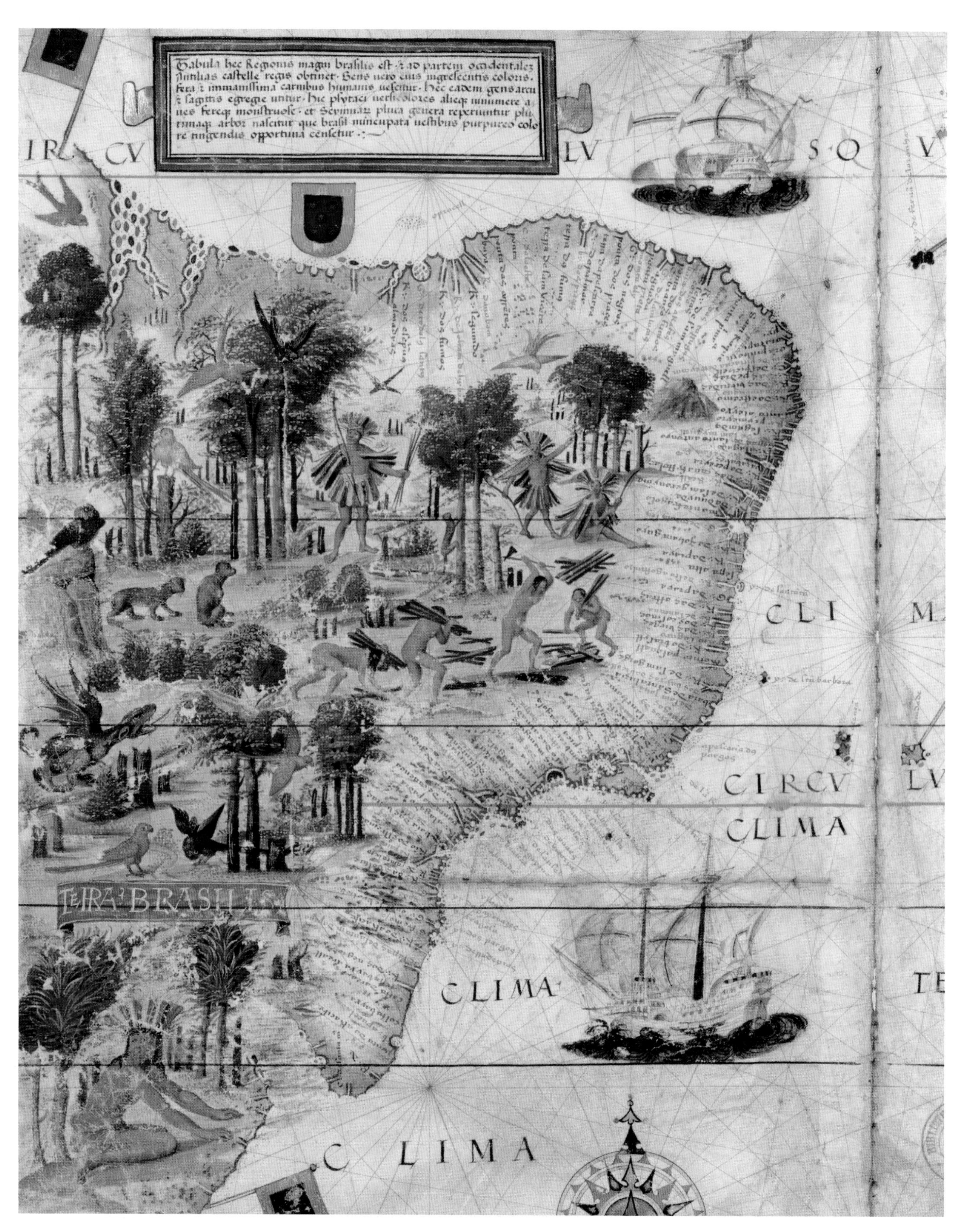

South America in the Lopo Homem-Reineis Atlas *shows the forest landscape together with the animals, birds and indigenous people found on the continent.*

PEDRO AND JORGE REINEL, LOPO HOMEM-REINEIS ATLAS, 1519

The age of exploration in the 16th and 17th centuries saw the world expand. Adventurers and invaders undertook mapmaking with enthusiasm and rigour, not least because mapping a territory was a significant help in claiming and retaining it. The *Lopo Homem-Reineis Atlas* (opposite and right) was produced in 1519 by two Afro-Portuguese cartographers, father-and-son team Pedro and Jorge Reinel. Their maps present natural features of a region, along with the names given to landmarks and settlements, useful information for shipping and for settlers. A world map shows an extensive southern continent, rumours of which had existed since antiquity, but the Reinels had no evidence on which to base their depiction of it.

The Reinels' fanciful depiction of a large southern continent was probably intended to dissuade Magellan from his circumnavigation of the globe by demonstrating that it would be impossible.

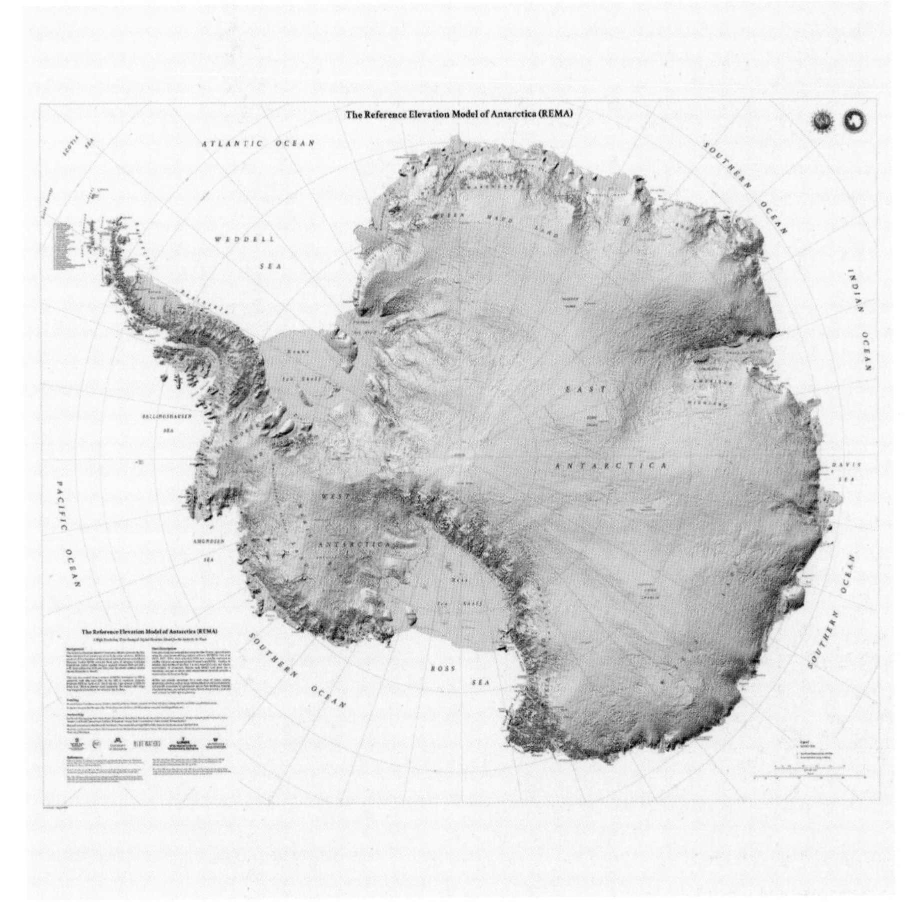

REMA, MAP OF ANTARCTICA, 2018

The fabled southern continent was eventually confirmed in 1820. Antarctica is much smaller than early theorists suggested – they thought there should be as much land below the equator as above it. Until 2018, it was less well mapped than Mars, but when the Reference Elevation Map of Antarctica (REMA) was completed, it suddenly became the best mapped continent on Earth.

The REMA was accomplished with the technology used to map other planets, including photographs taken by Earth-orbiting satellites during ten passes over the continent in 2015 and 2016. With no settlements, forests or rivers, it is very like the map of a distant planet or moon.

ATLANTIC OCEAN FLOOR
National Geographic Society
MELVIN M. PAYNE, PRESIDENT
for THE NATIONAL GEOGRAPHIC MAGAZINE
MELVILLE BELL GROSVENOR, EDITOR-IN-CHIEF; FREDERICK G. VOSBURGH, EDITOR
WILLIAM N. PALMSTROM, CHIEF, GEOGRAPHIC ART DIVISION
VERTICAL SCALE EXAGGERATED
GREENLAND
ICELAND
HUDSON BAY
CANADIAN SHIELD
NORTH AMERICA
SCANDINAVIA
EUROPE
BRITISH ISLES
NORTH SEA
BALTIC SEA
MID-OCEAN CANYON
SOHM ABYSSAL PLAIN
HATTERAS ABYSSAL PLAIN
BLAKE PLATEAU
BAHAMA
GULF OF MEXICO
PUERTO RICO TRENCH
CENTRAL AMERICA
IBERIAN PENINSULA
ATLAS MOUNTAINS
SAHARA
AFRICA
CAPE VERDE ISLANDS
AMAZON BASIN
SOUTH AMERICA
BRAZILIAN HIGHLANDS
CONGO BASIN
ANGOLA
KALAHARI DESERT
WALVIS RIDGE
RIO GRANDE RIDGE
SOUTH PACIFIC OCEAN
FALKLAND PLATEAU
FALKLAND ISLANDS
SCOTIA RIDGE
SOUTH SANDWICH TRENCH
INTERKART

THARP AND HEEZEN, LAND BENEATH THE SEA, 1977

We tend to think of Earth in terms of its landmasses as those are the most important areas to us, but they cover less than 30 per cent of the surface. Most of the Earth is underwater. Mapping the seabed was not possible until the middle of the 20th century. Today we use bathymetry to measure the contours of the seabed; this determines the depth of the water at intervals. The first method used was sonar, bouncing sound off the seabed and measuring the time taken for its echo to return. It was a slow process that could be carried out only by crossing the ocean, taking repeated readings. Later, lasers and satellite altimetry were used, as they are to map other planets. The first three-dimensional mapping of the ocean floor was carried out in 1957 by Marie Tharp and Bruce Heezen, and the first digital map of the world's oceans was created in 1970.

This detailed image of the seabed northeast of Saunders Island in the southern Atlantic Ocean shows how sediment is flowing down deep canyons from the volcanic island into the sea. Areas of shallow sea are red, and areas of deep sea are purple. The map was made in 2010 by the British Antarctic Survey.

Facing page: *Tharp and Heezen's map, created in 1977, shows the Mid-Atlantic Ridge running along the floor of the Atlantic Ocean. It's a vast underwater mountain range produced by hot, semi-liquid rock (magma) leaking up from below as the tectonic plates that carry Earth's crust move apart.*

THE SURFACE AND BELOW

Maps of Earth take many forms, including contour maps that reveal the topography of the land, geological maps showing its composition, and magnetic and gravitational maps showing the distribution of Earth's magnetic field and density. Since the 1970s, plate tectonics theory has explained how the continental landmasses sit on large 'plates' of the Earth's crust which are in constant motion, carried by slowly moving magma (thick, molten rock). Volcanic eruptions and earthquakes are evidence of tectonic activity, as is the rearrangement of the land masses which group into supercontinents over hundreds of millions of years and then break up and disperse. Earth is the only planet we can map through time.

WILLIAM SMITH, GEOLOGY OF ENGLAND AND WALES, 1815

William Smith's geological map of England and Wales, published in 1815, laid the foundations of modern geology. The Industrial Revolution created a great demand for coal. Smith worked as a surveyor, inspecting coal mines, and began to map the strata of rocks of different types and ages, eventually mapping all of England and Wales.

Our current understanding of Earth's geological activity – tectonics and the movement of landmasses – relies on tracing the age and composition of rocks. Smith calculated their age by the fossils they contained. Today, the presence of radioactive isotopes is often a more accurate way of assessing their age.

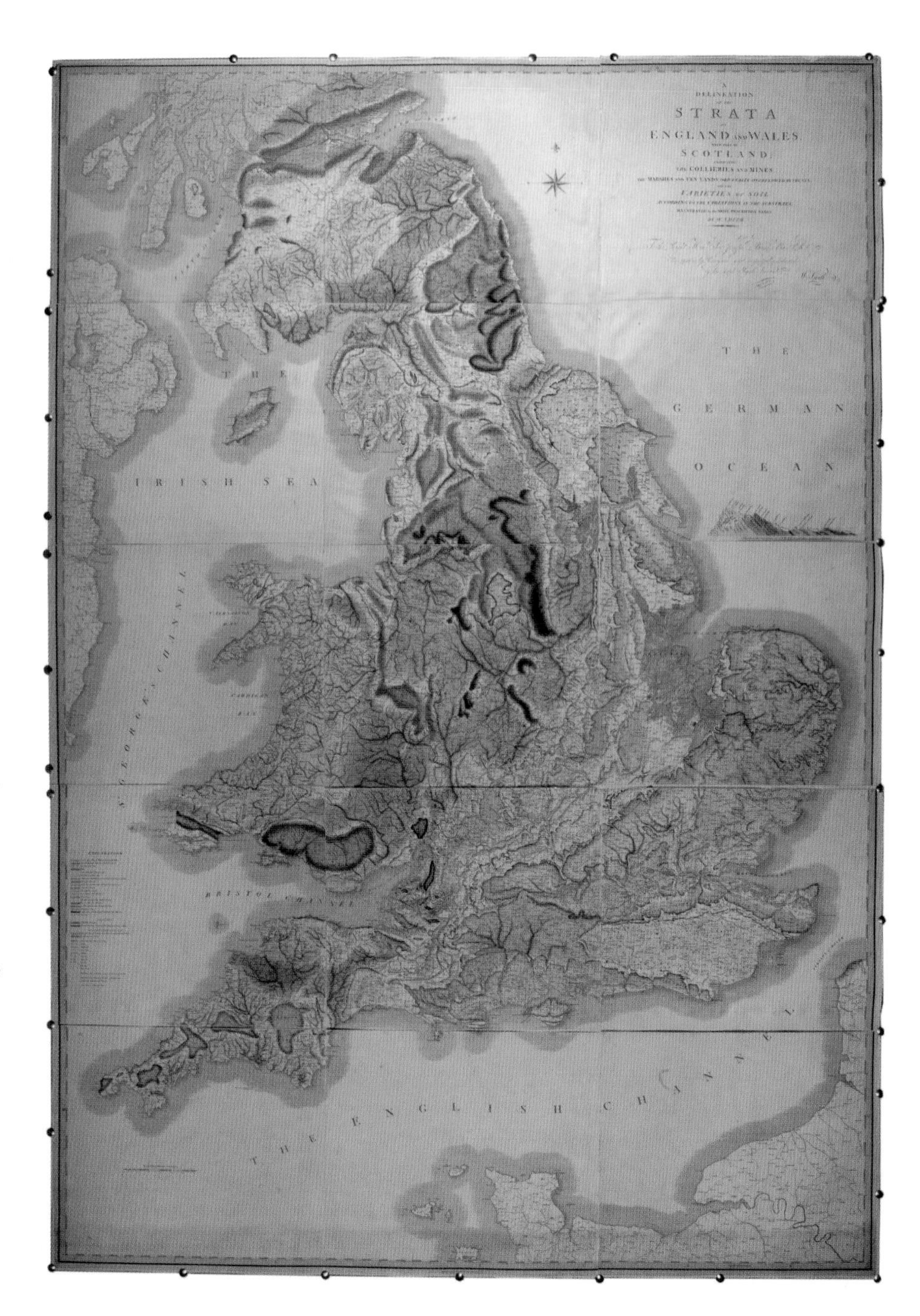

The Earth's crust is divided into tectonic plates which move slowly on top of the flowing magma. The boundaries between plates (marked in red on the map above) are sites of earthquake and volcanic activity, including the mid-ocean ridges.

ANTONIO SNIDER-PELLEGRINI, CONTINENTS FITTING TOGETHER, 1858

Long before the movement of tectonic plates was understood, it was noticed that the coasts of West Africa and South America mirror each other. Snider-Pellegrini explained this by suggesting that the continents were created by God like this but separated by a vast volcanic eruption on the sixth day after Creation.

OUR PLANET AS ALIENS SEE IT

Our close-up view of Earth has been the only one possible for almost all of human history. With the advent of powered flight, it finally became possible to see the land from above. But even from a plane, in a cloudless sky, the horizon is 391 km (243 miles) away, making it possible to view perhaps a couple of American states or half a smallish country. Viewing the whole Earth, in the way we can view the other planets, only became possible with space travel.

APOLLO, EARTHRISE, 1968

This photo of Earth rising above the surface of the Moon was taken by the crew of Apollo 8 in 1968. Originally in black and white, it was remastered in 2018 to add colour. If aliens came close enough – orbiting the Earth as our satellites do – they would have some spectacular views of our planet.

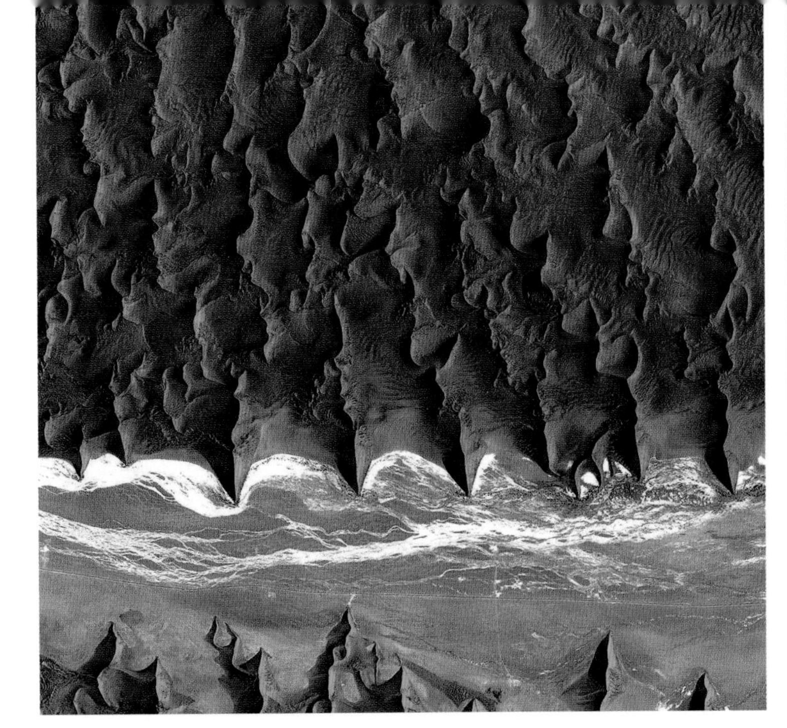

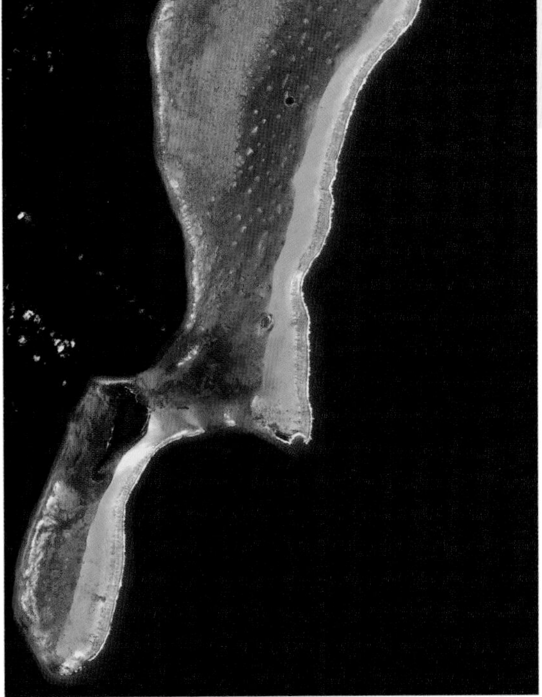

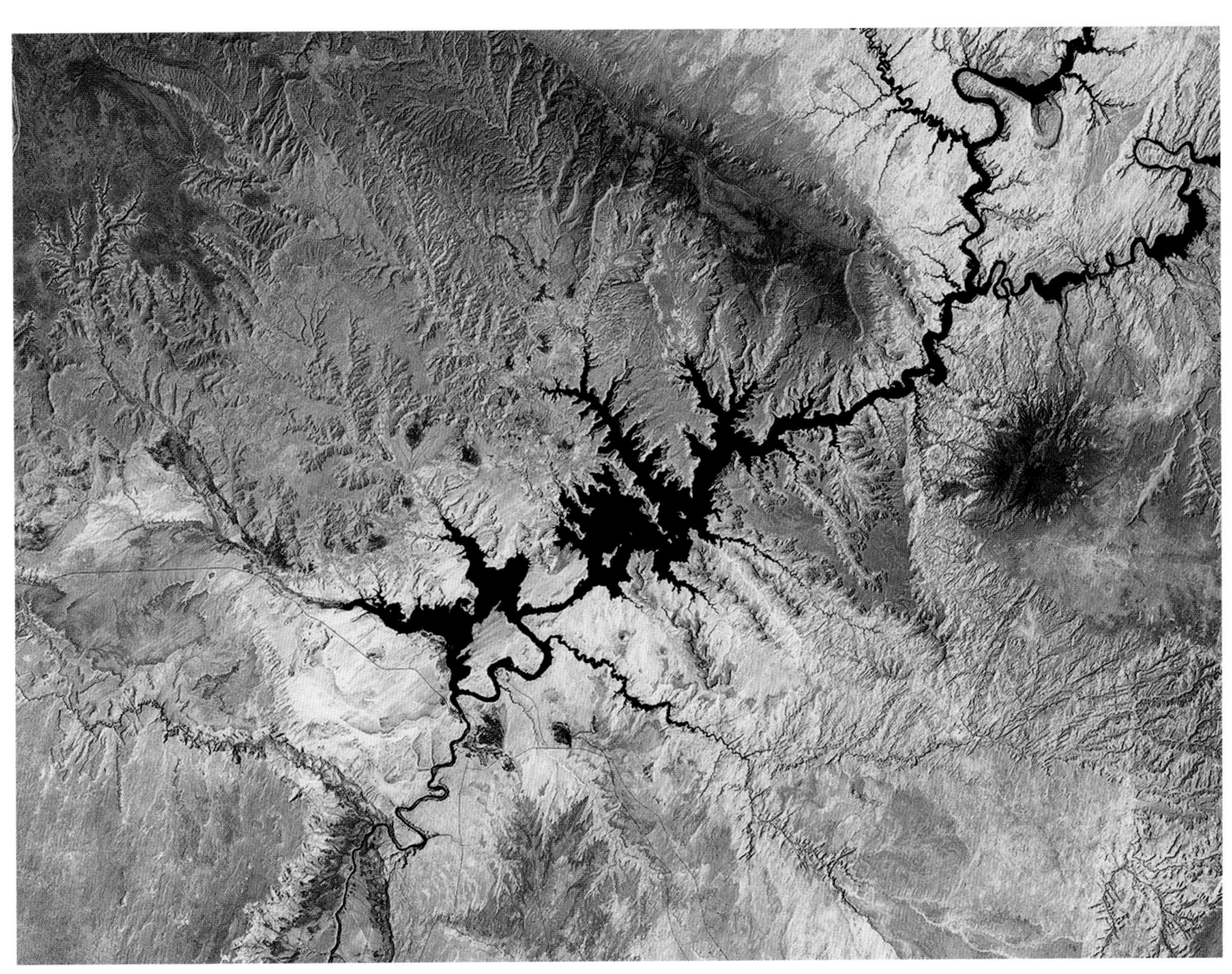

The Namib Desert (top left), Kompsat-2 satellite, 2012: the blue area is the dry river bed of the Tsauchab, with bright white deposits of salt. Lighthouse Atoll, Belize (top right), ALOS satellite, 2011: the Great Blue Hole appears as a dark circle; it was formed when the area was dry land, before it flooded at the end of the last ice age. Lake Powell on the Colorado River, USA (above), Landsat 2011: a false-colour image of the region.

This photograph of the Moon beginning a transit across the face of the Earth was taken by NASA's DSCOVR satellite in 2016. DSCOVR is in orbit around the Sun, monitoring from a distance of 1.5 million km (930,000 miles).

EARTH'S MOON

Given that neither Mercury nor Venus has a moon, ours is the closest moon to the Sun. It was probably formed when a planet around the size of Mars crashed into Earth 4.5 billion years ago. A vaporized mix of the planet and Earth settled back onto a largely molten Earth and made a cloud of debris which coalesced to form the Moon. The Moon's composition is therefore very similar to that of Earth.

A STORMY START

In the early days of the solar system, rocky and icy bodies frequently collided with the inner planets and their moons. These were leftovers from the formation of the planets. One theory suggests a period called Late Heavy Bombardment (LHB) around 4.1–3.8 billion years ago represented a new assault, when the inner planets were pummelled relentlessly by lumps of rock and ice after the initial chaos had died down. Evidence for the LHB is far from conclusive. The

craters on the Moon are testament to the battering our satellite has taken since its formation. It's likely that collisions were initially intense, but gradually reduced. Earth would have been similarly assaulted, but a combination of wind, water and tectonic activity has obliterated the evidence. On the Moon, where there is little weathering and no moving tectonic plates, the craters have piled up, one within another, for billions of years.

Crater Daedalus, on the far side of the Moon, photographed by Apollo 11 in 1969. The crater, about 80 km (50 miles) across, is surrounded by smaller ones, evidence of many other impacts.

GALILEO, THE IMPERFECT MOON, 1609

Although there are quite clearly surface features on the Moon, very few depictions of them survive from before the invention of the telescope. In the West, it was commonly believed that the heavens are perfect and uniform. Dark patches on the Moon were explained away as areas of different density that allowed through different amounts of light. This meant they were not really features and didn't merit noting and mapping. (Now astronomers are interested in the variable density of different parts of a planet or moon and make gravity maps.)

With the advent of the telescope, this view became untenable. Galileo made some of the first sketches of the Moon through a telescope. He dared to describe the features of the Moon as mountains and valleys or dips, just as we find on Earth, with sunlight illuminating one side and casting the other side into shadow. This was provocative as it contradicted the prevailing model of a God-made, perfect surface. A group of Jesuits set to examine Galileo's findings decided he was wrong: the Moon, they maintained, is perfectly even.

FRANCESCO FONTANA, PHASES OF THE MOON, 1646

Francesco Fontana was an Italian maker of telescopes. He was the first person to use convex lenses, and made the most accurate drawings of the Moon's surface at the time. These were distributed around Europe without acknowledging his authorship, a slight which prompted him to produce a book of observations, *New Celestial and Terrestrial Observations*, published in 1646. This included some of his earlier drawings and contained the first accurate atlas of the Moon, showing its phases in twenty-seven images.

Left: *The telescope in this detail from* The Allegory of Sight *by Jan Brueghel the Elder and Peter Paul Rubens was almost certainly made by Fontana, as he was the only person known to be making silver telescopes like this at the time.*

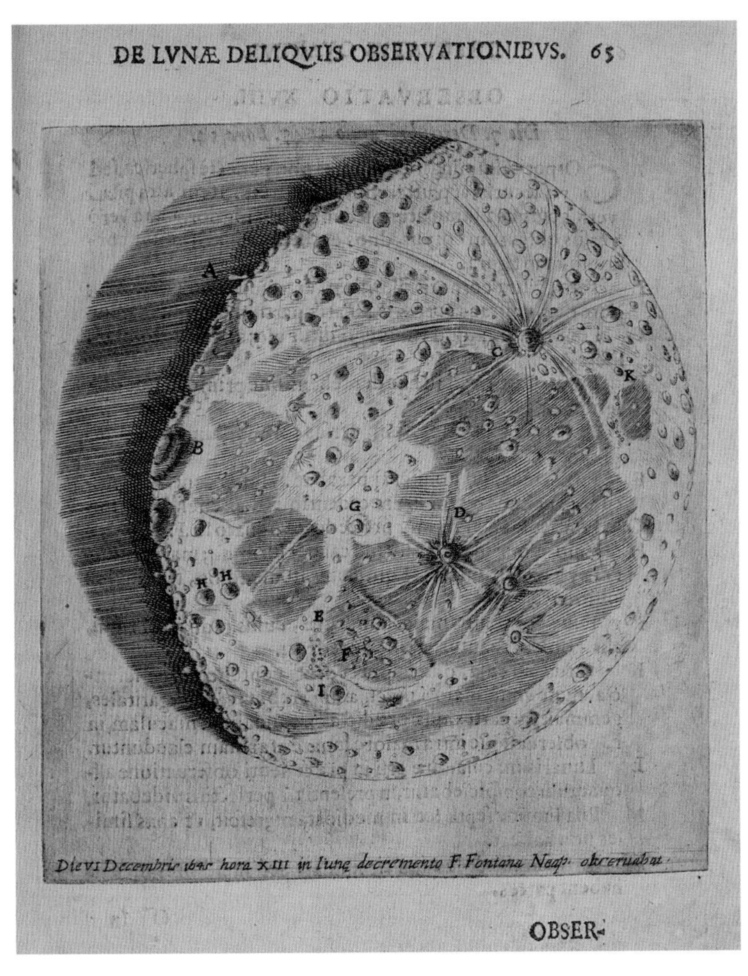

Francesco Fontana's drawing of the Moon (left), made in 1629, and a photo taken from the Apollo 11 spacecraft (above). South is at the top in both images.

CLAUDE MELLAN, MOON, 1635

Less than 30 years after Galileo's sketches, astronomer Pierre Gassendi and humanist Nicolas-Claude Fabri de Peiresc commissioned three images of the Moon from French engraver Claude Mellan. These show considerable improvement in the power of the telescope and in the detail and accuracy of observations of the Moon's surface. In these engravings, the lines radiating from relatively new craters are clearly visible – a detail that Galileo could not see. The image on the right shows the full Moon (the other two show the Moon in its first and third quarters).

MICHAEL VAN LANGREN, NAMING FEATURES, 1645

Dutch astronomer Michael van Langren was keen to produce an accurate map of the Moon for a practical purpose: to aid sailors with navigation, enabling them to determine longitude on any day. (This would have taken 30 maps, which he did not complete.) As there were no conventions for naming places anywhere other than Earth, Langren had a free hand and named most major features after Spanish royalty and Catholic saints. Outside Spain, and among Protestants, these were not popular and were soon dropped, but those features that van Langren had named for astronomers, mathematicians and other intellectuals were retained.

JOHANNES HEVELIUS, SELENOGRAPHIA, 1647

Another person to name places on the Moon was the Polish amateur astronomer Johannes Hevelius. In his atlas of the Moon, *Selenographia*, he used a consistent scheme of naming raised areas *mons* (mountain) and large depressed areas *mare* (sea). Small depressions which he could see only through the telescope he designated craters. He named 268 features, ten of which still go by the names he gave them. This version of Hevelius's map was published in 1696 by Daniel Zahn.

J. W. DRAPER, EARLY PHOTOGRAPH OF THE MOON, 1840

Over the years, drawings of the Moon became increasingly accurate, but in 1827 the advent of photography marked a new departure. It was some decades before techniques improved sufficiently to provide photographs of the Moon that surpassed the drawings of experienced astronomical draughtsmen. The American astronomer J. W. Draper took his first photograph of the Moon in 1839. This better image is from 1840.

Draper's image of the Moon was taken using the daguerrotype process introduced by artist and photographer Louis Daguerre in 1839.

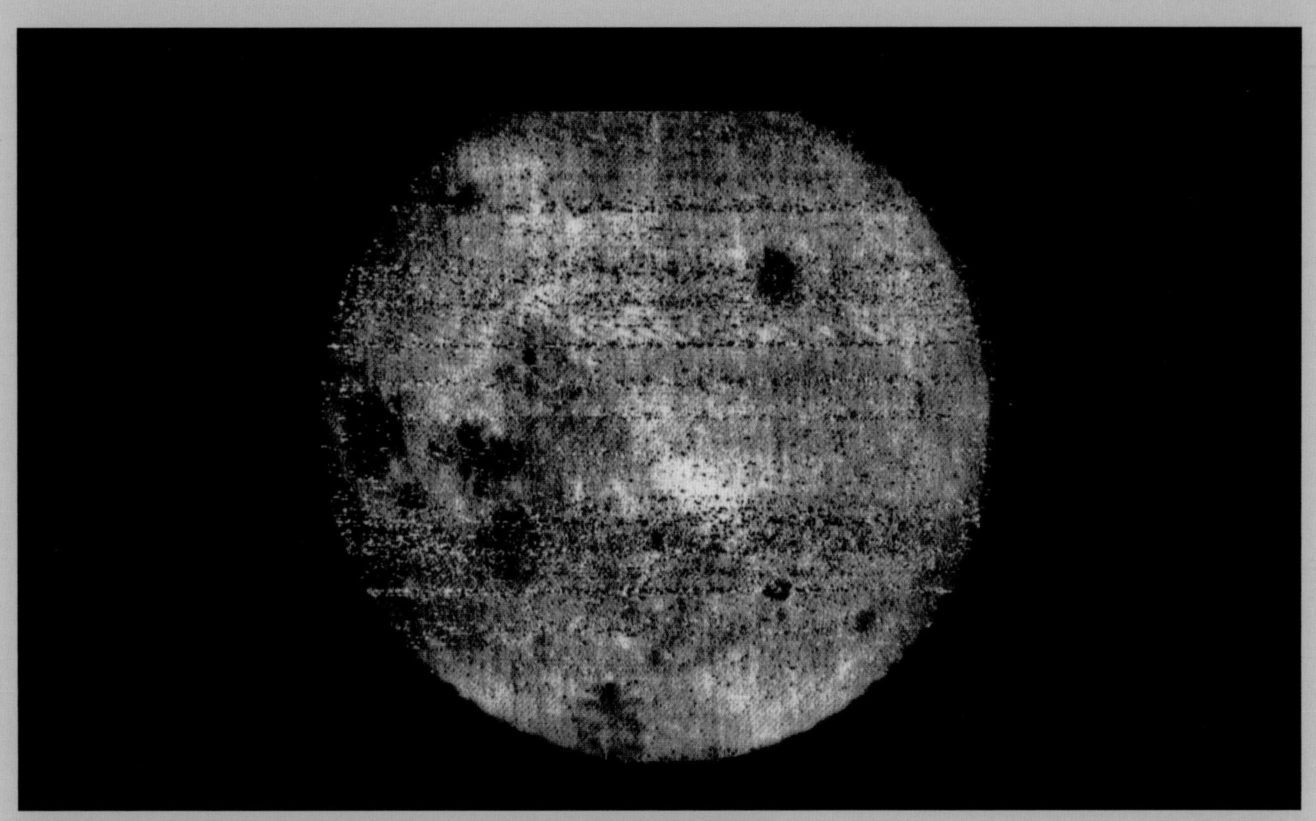

NASA, FAR SIDE OF THE MOON, 2009

Before 1959, no one had ever seen the far side of the Moon. This is because the Moon is tidally locked to the Earth, so the same side is always facing us. The Soviet spacecraft Luna 3 was the first to transmit photographs from the far side, orbiting the Moon in 1959. It was quickly apparent that we get to see the more interesting side. The far side is extravagantly cratered, but has few *maria*, the large, dark plains of basalt that occupy nearly a third of the surface of the near side. The image on the left, produced from data collected by NASA's Lunar Reconnaissance Orbiter in 2009, shows the highly cratered surface.

Top: *Humankind's first ever view of the far side of the Moon from 1959.*

Bottom: *The far side of the Moon from data collected in 2009. The prominent dark spot upper right is* Mare Moscoviense, *one of few* maria *on the far side.*

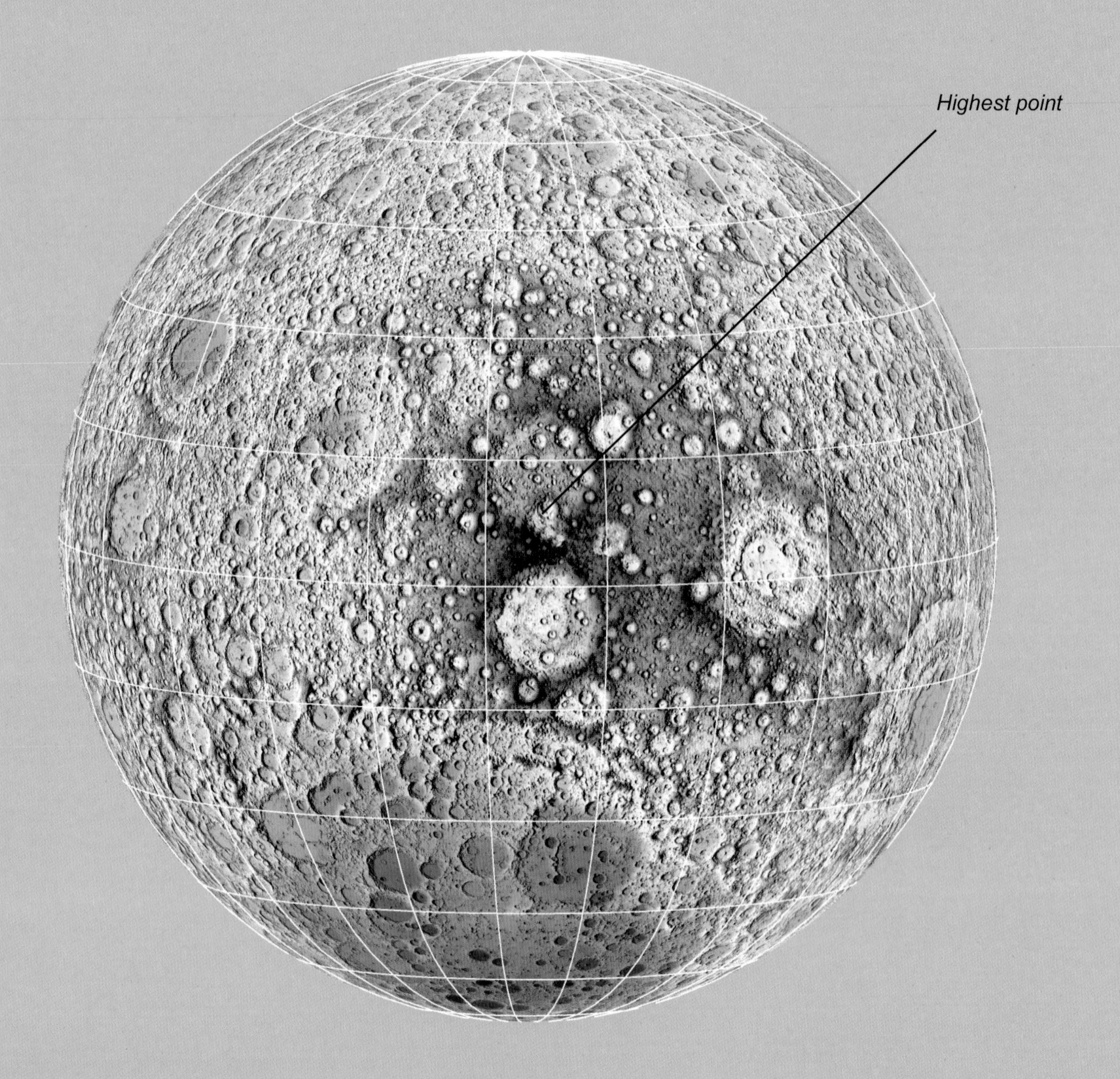

UNITED STATES GEOLOGICAL SURVEY, THE MOON'S TOPOGRAPHY, 2015

The USGS map of the topography of the Moon was compiled using more than 6.5 billion readings collected by NASA's Lunar Reconnaissance Orbiter (LRO) between 2009 and 2013. The LRO's mission was to map the surface of the Moon in greater detail than ever before. It took high resolution photographs and returned details of topography, roughness of the surface, slopes and other features. One aim of the mission was to identify locations suitable for landing and to provide data to facilitate future missions.

Facing page, bottom: Mare Orientale, *left of centre, is a large impact crater partially flooded by balsatic lava. The crater is 950 km (600 miles) across; the concentric circles around it were formed by ripples in the Moon's crust resulting from the impact.*

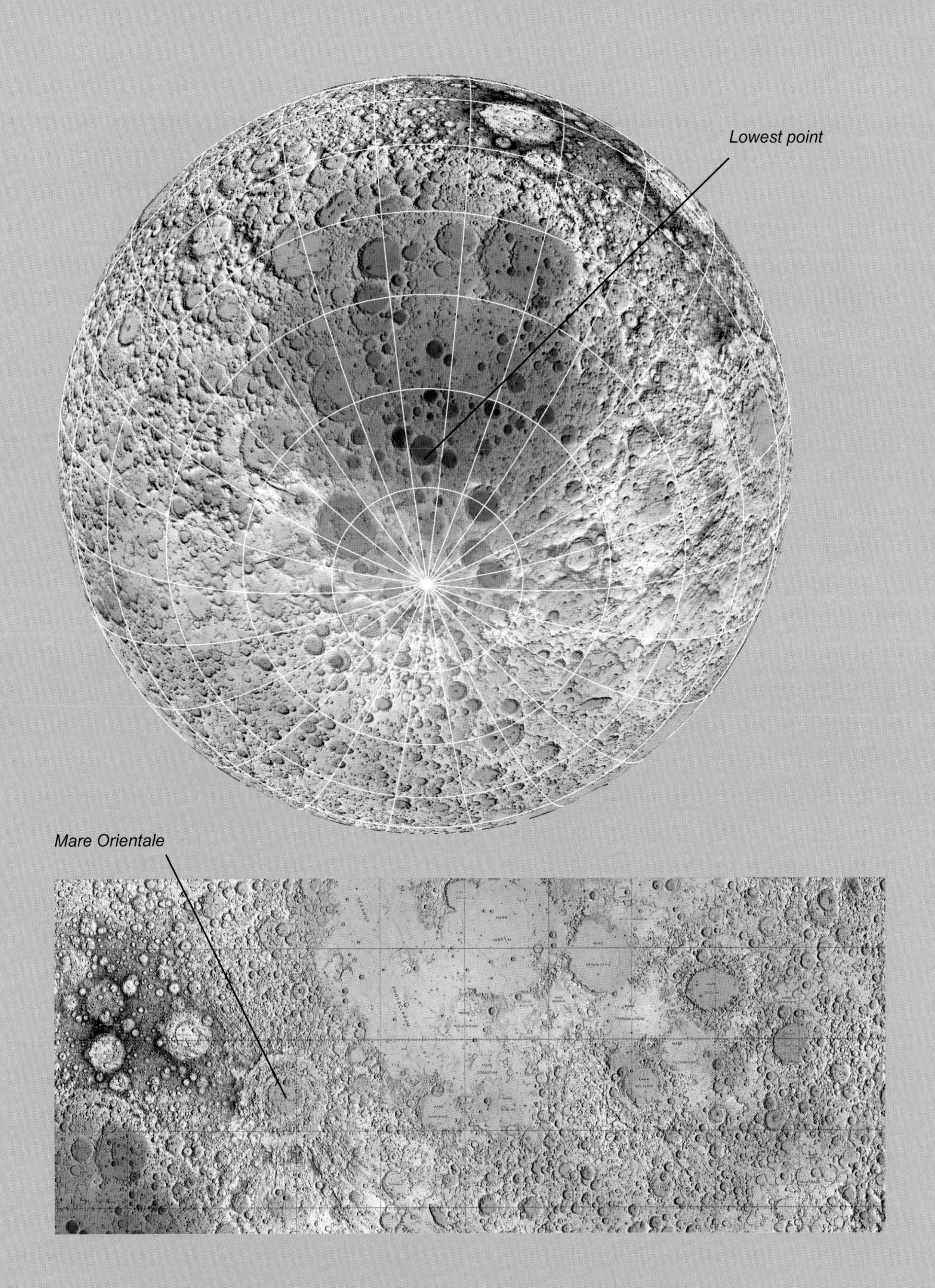
Lowest point
Mare Orientale

NASA, GRAVITY PROFILE OF THE MOON, 2012

NASA's Gravity Recovery and Interior Laboratory (GRAIL) mission consisted of twinned spacecraft called Ebb and Flow orbiting the Moon and mapping its gravity using microwave radiation.

The strength of the gravitational field varies over the surface, depending not just on the distance to the core at any particular point but on the composition of the rocks beneath the surface. These gravity maps of the Moon reveal the underlying gravity after the surface topology has been removed from consideration – so representing only the density of the subsurface. The gravity profile reveals that although the far side of the Moon has more visible craters, both sides have been bombarded equally by meteors. The near side is hotter with a thinner crust; this is believed to be because there is a higher proportion of heat-producing radioactive elements on the near side. The higher temperature means that an impactor creates a larger crater on the near side than it would initially create on the far side, but these craters have filled with volcanic lava, now hardened, hiding the base of the crater.

In 2019, scientists analyzing gravity maps suggested that an area of high gravity under the Aitken basin at the Moon's south pole (the large red circle in the central image above) represents a dense meteor embedded beneath the surface.

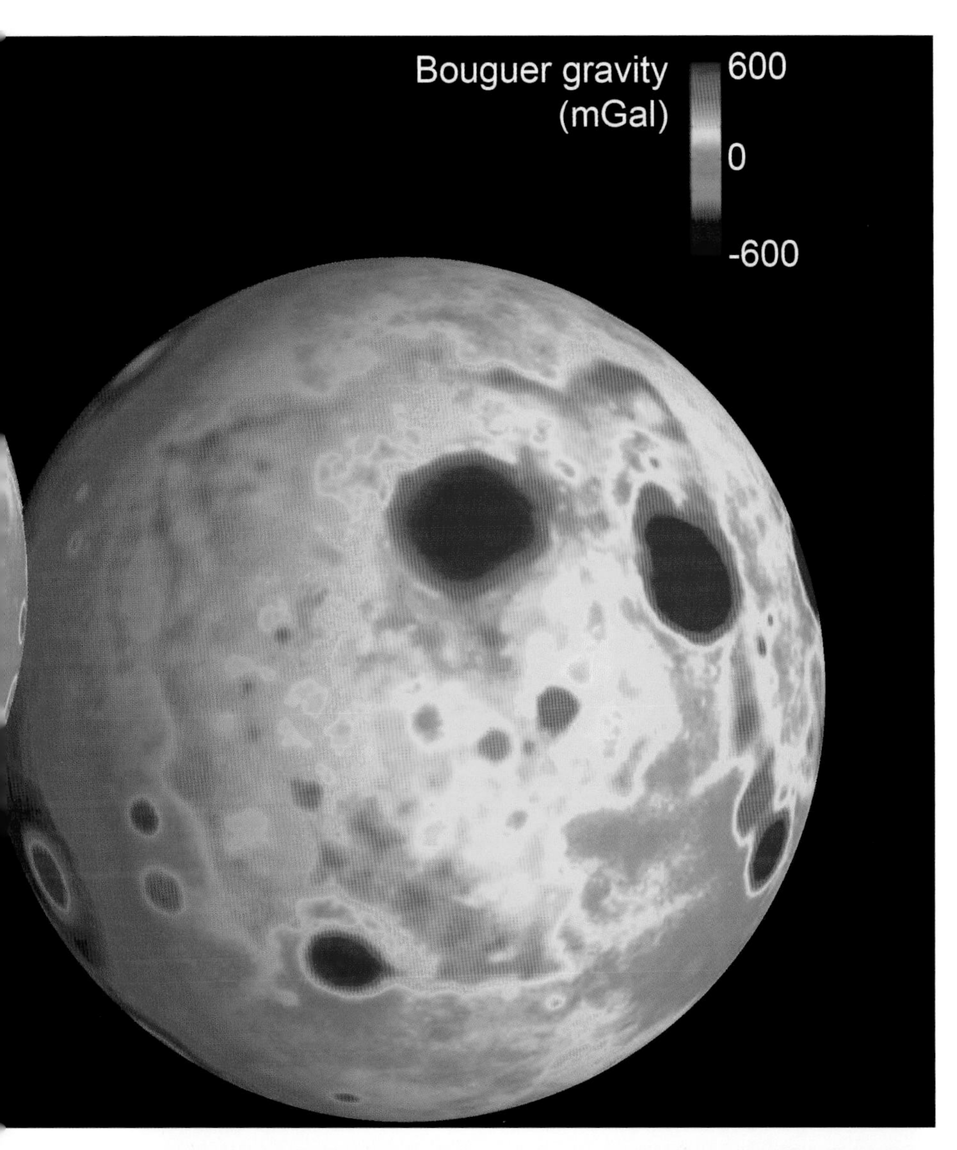

NASA AND OTHERS, GRAVITY GRADIENTS AROUND *OCEANUS PROCELLARUM*, 2018

This image of the near side of the Moon superimposes gravity data (in blue) over the topographical map of the Moon derived from the Lunar Reconnaissance Orbiter data. The deep blue lines are thought to indicate ancient rift zones flooded with lava, now buried beneath the volcanic plain.

CLEMENTINE, COMPOSITION OF THE MOON, 1994

The Clementine mission, launched in 1994, spent two months mapping the Moon at eleven different wavelengths, collecting data about topography, gravity and chemical composition.

This map shows the distribution of the chemical element iron in the surface of the Moon. A high level is found in the *maria* of the nearside and in the south pole-Aitken basin on the far side, the largest impact crater in the solar system. (Grey areas are unmapped.) In the Aitken basin, lower level crust has been exposed by a meteor gouging out the surface. The map supports the theory that rocks on the Moon become more mafic (rich in magnesium and iron) with depth.

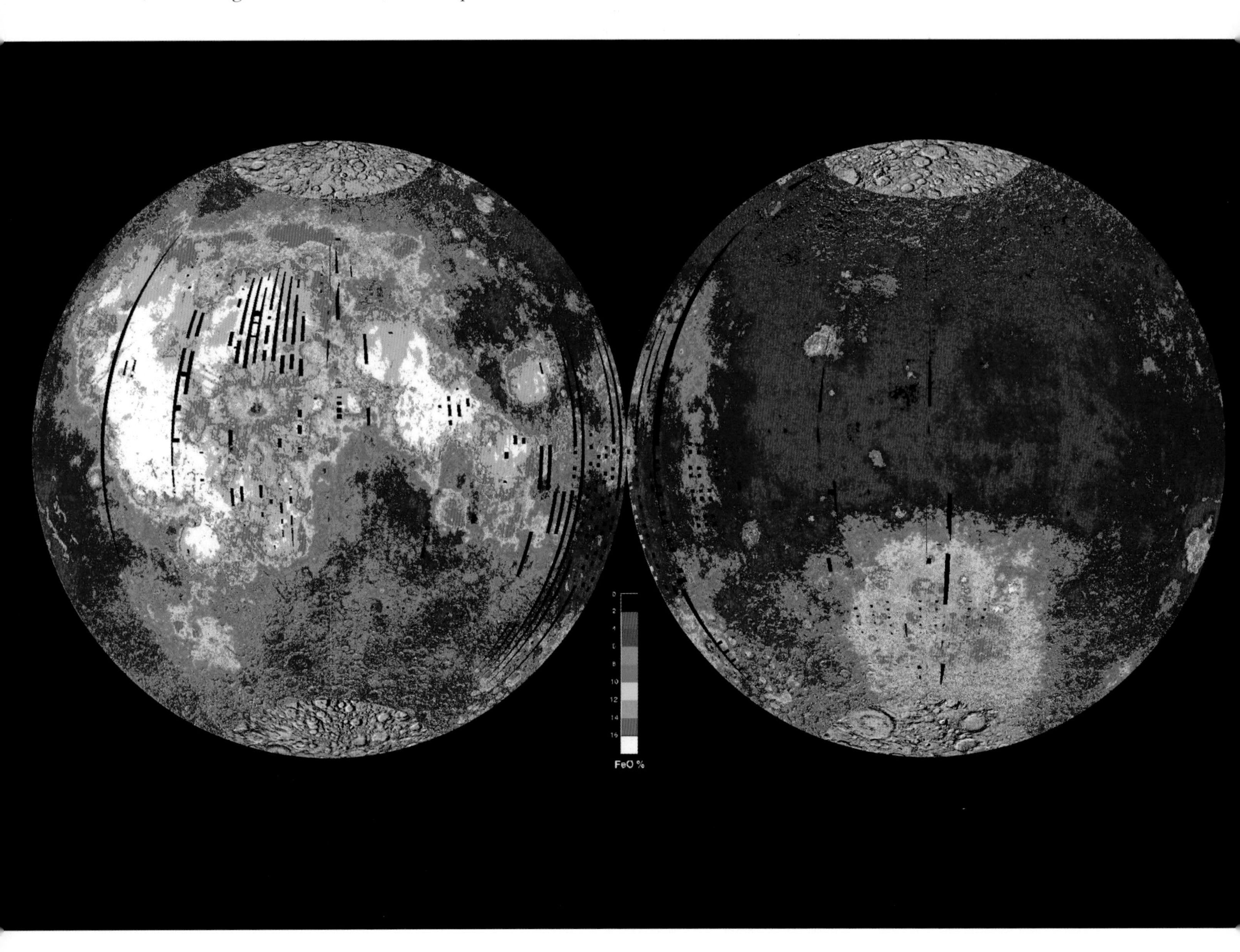

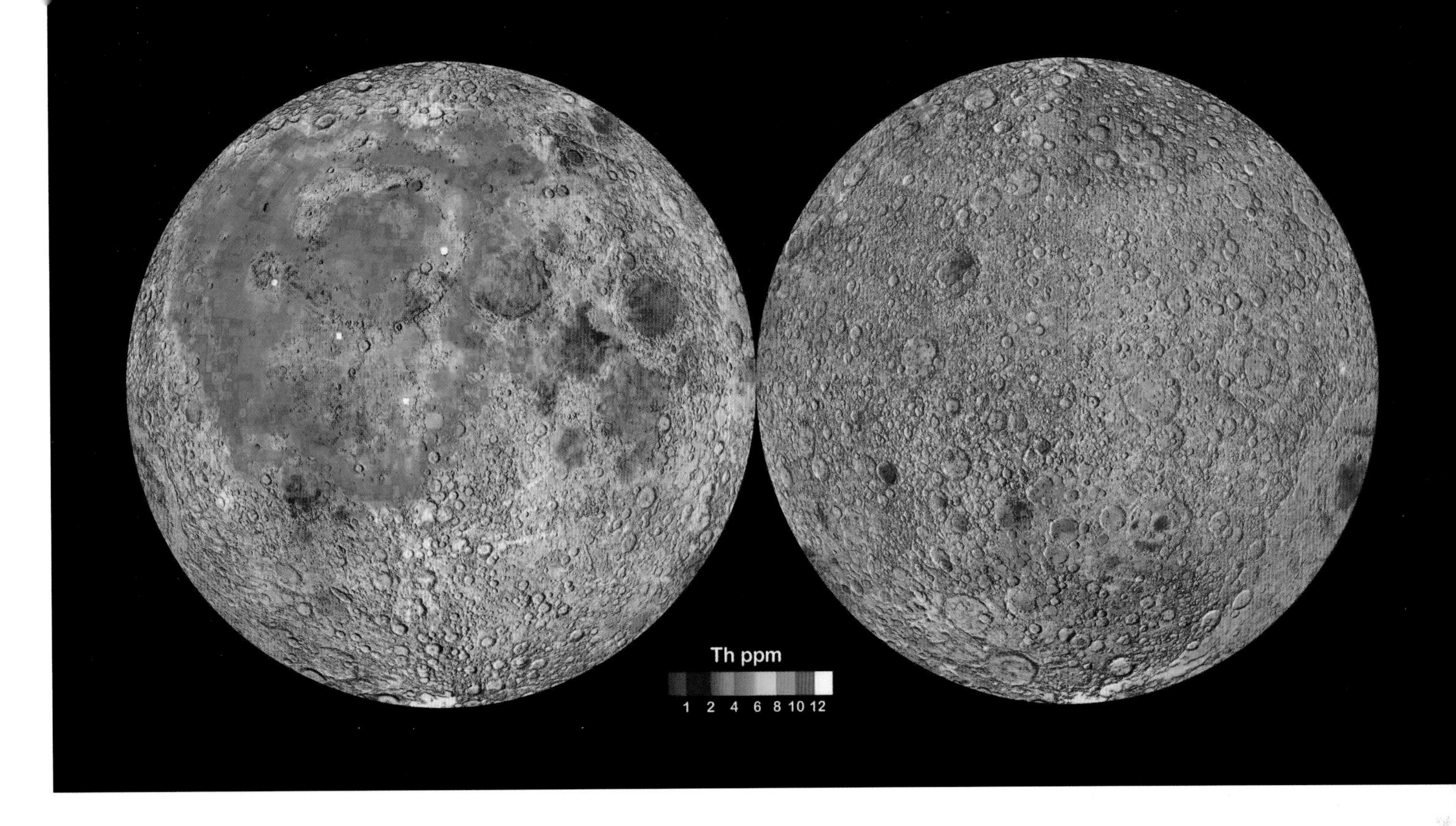

NASA'S LUNAR PROSPECTOR, THORIUM CONCENTRATION, 2006

The Lunar Prospector mission of 1998–9 collected data on the distribution of a group of elements important in the Moon's crust, including uranium and thorium. After its formation, when the Moon was largely molten, these elements were among the last to solidify. Before the Prospector mission, scientists expected thorium to be equally distributed across the Moon, but it is massively concentrated in the area of the Imbrium basin and *Oceanus Procellarum*, with a smaller concentration in the south pole-Aitken basin. This anomaly has not yet been explained.

USGS, MOON LANDING SITES

The Moon has been visited by numerous missions – by NASA, the Soviet space agency and, most recently, the Chinese National Space Administration. This means there are now sites on the Moon that have something other than natural geological features. The landers and equipment taken to the Moon are largely still there, as is the debris of craft which have been deliberately crashed there. Knowing the location of these sites is important for future landings and impacts. The lack of atmosphere or moving water on the Moon means that the objects left there will remain, unless destroyed by impacting meteors or spacecraft, for millions or billions of years.

MARS

Mars has been the focus of human imagination since the 19th century, sparked by being (usually) our closest neighbour and by the belief that canals were visible on the surface, suggesting intelligent life. The possibility of other living things sharing our solar system has both captivated and terrified us. Martians have become a staple of science fiction and we haven't yet ruled out finding some kind of microorganisms on Mars or evidence of their existence in the past. In the 19th century, people were so convinced that Mars was Earth's

A global mosaic made from images taken by NASA's Viking 1 Orbiter in 1980.

planetary twin that the French astronomer and author Camille Flammarion could write in 1896: 'The world of Mars is so much like the world of Earth that, had we travelled thither someday and forgotten our route, it would be almost impossible for us to tell which of the two is our native planet. Without the Moon, which would mercifully relieve our incertitude, we would run the enormous risk of calling upon the natives of Mars while assuming we have landed in Europe or in some terrestrial neighbourhood.'

Known as the 'red planet' even before the days of the telescope, Mars' reddish tinge is caused by the high level of iron oxide (rust) in its surface. It is now a cold dry world; in the past it seems to have had water, but today this is locked up in ice and held in clouds in its thin atmosphere of carbon dioxide.

Planetary focus	Mars
Length of year	687 days
Length of day	24.5 hours
Size x Earth mass	0.1
Size x Earth radius	0.5
Average distance from Sun	229 million km (142 million miles)
Discovered	Prehistory

A crater 42 km (26 miles) across lies near the equator, on the rim of the vast Schiaparelli crater on Mars.

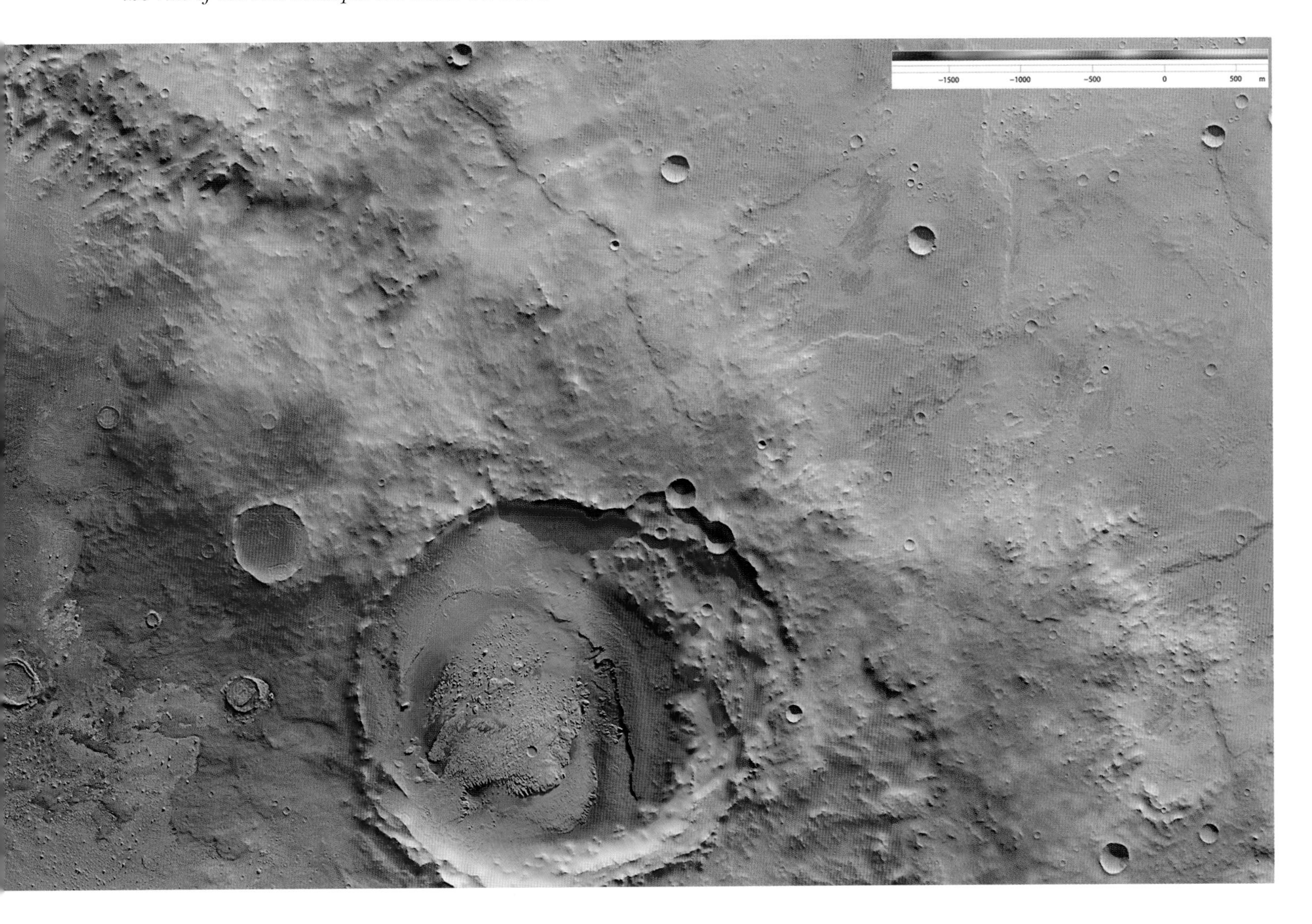

BRINGING MARS INTO FOCUS

When Galileo first looked at Mars through his telescope, he couldn't make out any surface features. It took better telescopes than his to resolve details on the planet and, even then, little was visible at first. Fontana made the first sketches of Mars in 1636 and they show nothing recognizable. Huygens made out a shady patch in 1659 (see page 20). Both Robert Hooke in England and Giovanni Cassini in Italy drew the planet in 1666, finding more variation in shading, and Cassini seems to have noticed the polar ice caps.

Unlike the Moon, which always shows us the same face, Mars quite clearly rotates and astronomers see different parts of the planet. As the Martian day is only slightly longer than an Earth day (by about 37 minutes), an astronomer viewing the planet at the same time each day will see it change its aspect slowly over a short period of time.

Above: *Cassini's sketch of the surface of Mars shows the variations he could make out.*

Below: *Kepler's diagrams of the orbit of Mars.*

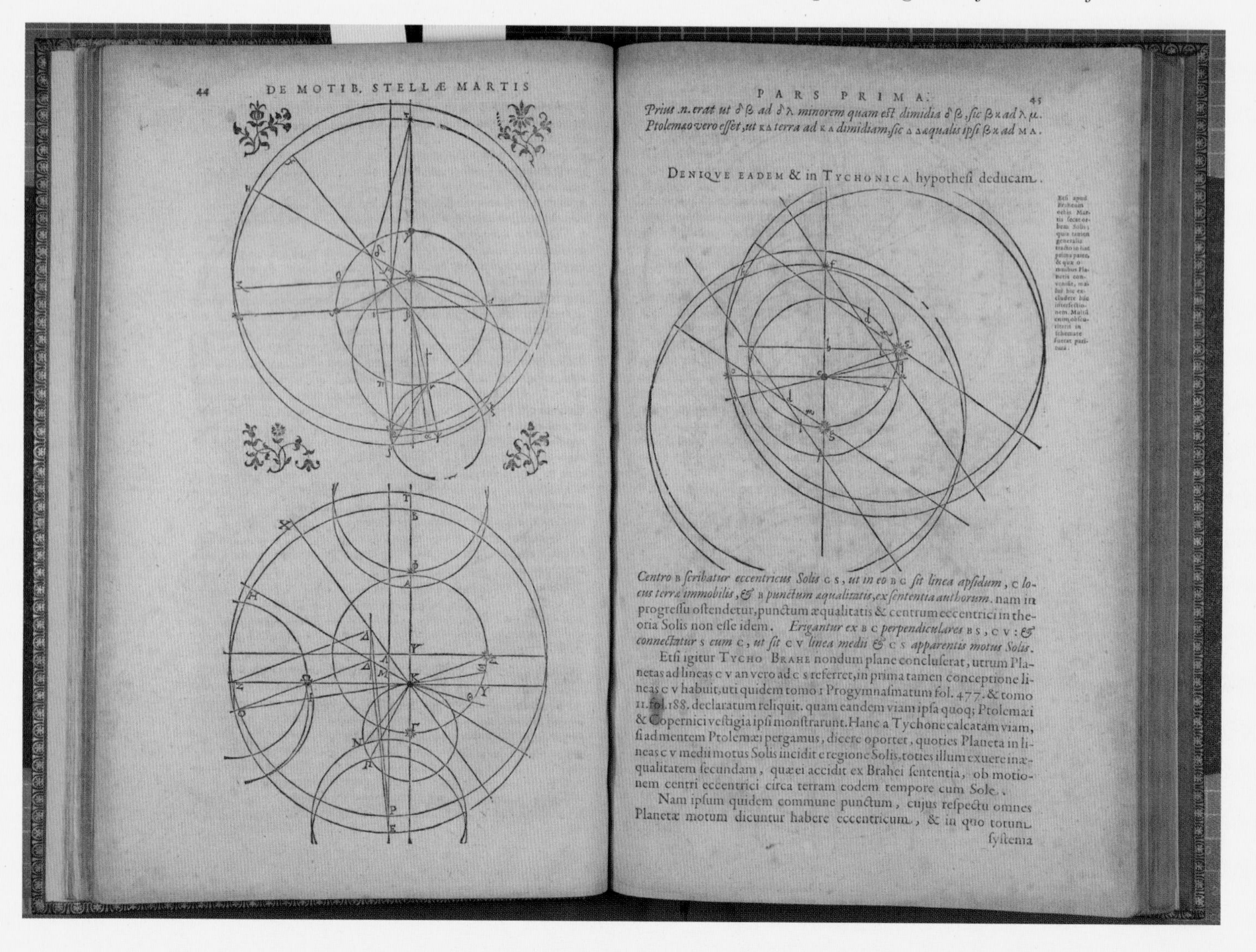

44 DE MOTIB. STELLÆ MARTIS

PARS PRIMA. 45

Prius .n. erat ut δβ ad δλ minorem quam est dimidia δβ, sic βκ ad λμ. Ptolemæo vero esset, ut KΔ *terra ad* KΔ *dimidiam, sic* ΔΔ *æqualis ipsi* βκ *ad* MΔ.

DENIQUE EADEM & in TYCHONICA hypothesi deducam.

Centro B *scribatur eccentricus Solis* GS, *ut in eo* BG *sit linea apsidum*, C *locus terræ immobilis, &* B *punctum æqualitatis, ex sententia authorum*. nam in progressu ostendetur, punctum æqualitatis & centrum eccentrici in theoria Solis non esse idem. *Erigantur ex* BC *perpendiculares* BS, CV: *& connectatur* S *cum* C, *ut sit* CV *linea medii &* CS *apparentis motus Solis*.

Etsi igitur TYCHO BRAHE nondum plane concluserat, utrum Planetas ad lineas CV an vero ad CS referret, in prima tamen conceptione lineas CV habuit, uti quidem tomo I Progymnasmatum fol. 477. & tomo II. fol. 188. declaratum reliquit. quam eandem viam ipsa quoq; Ptolemæi & Copernici vestigia ipsi monstrarunt. Hanc a Tychone calcatam viam, si ad mentem Ptolemæi pergamus, dicere oportet, quoties Planeta in lineas CV medii motus Solis incidit e regione Solis, toties illum exuere inæqualitatem secundam, quæ ei accidit ex Brahei sententia, ob motionem centri eccentrici circa terram eodem tempore cum Sole.

Nam ipsum quidem commune punctum, cujus respectu omnes Planetæ motum dicuntur habere eccentricum, & in quo totum systema

MEASURING FROM MARS

Although astronomers had a good sense of the relative distances between the planets, they had no way of judging the absolute distances. This changed when Cassini, enlisting the help of fellow astronomer Jean Richer, measured the parallax of Mars.

Cassini remained in Paris while Richer went to French New Guinea. By comparing the apparent position of Mars relative to the stars behind them in both places, Cassini could calculate the distance from Earth to Mars. He found it to be 140 million km (86,991,967 miles), just 7 per cent less than the figure accepted today of 150 million km (93,205,679 miles).

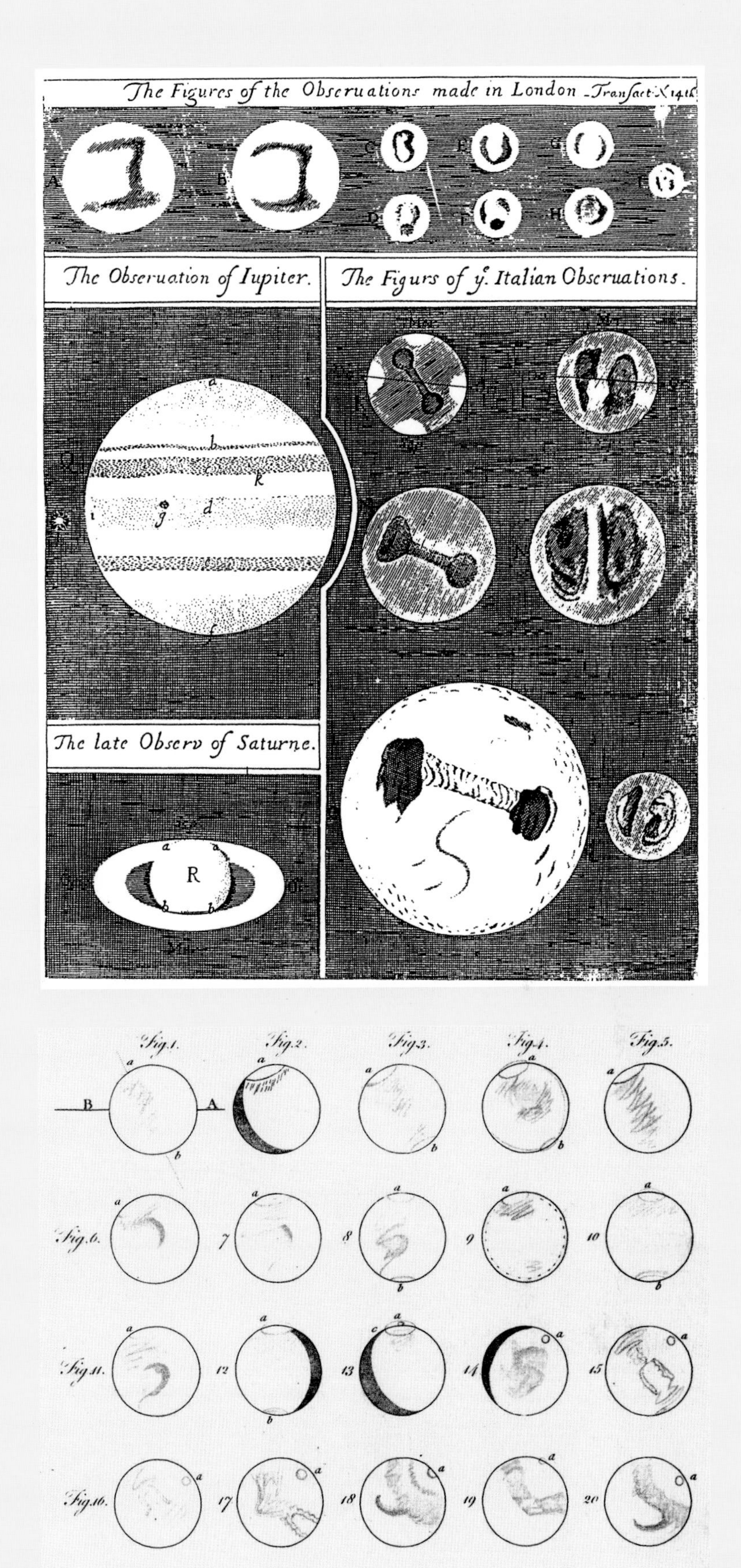

Right: *Cassini's drawings of the features of Mars. He was the first to mention the polar ice caps. He used his observation of the surface to measure Mars' orbital period, finding a day-length of 24 hours and 40 minutes, less than three minutes different from the value now accepted.*

Right: *An additional issue for early astronomers with telescopes was that Mars is sometimes very far from Earth. Surface details would not have been visible when Mars was too far away. Herschel's drawings, made in 1783, capture surface shading and often a polar ice cap.*

JOHANN MÄDLER AND WILHELM BEER, MAP OF MARS, 1840

Two German astronomers, Johann Mädler and Wilhelm Beer, made the first true map of Mars in 1831 and then quickly made several others. They plotted what they decided were probably permanent features of Mars (rather than clouds or atmospheric phenomena) and constructed a global coordinate system which is still used today. Although their maps were essentially still patterns of light and shade, the patches can today be mapped to recognized features.

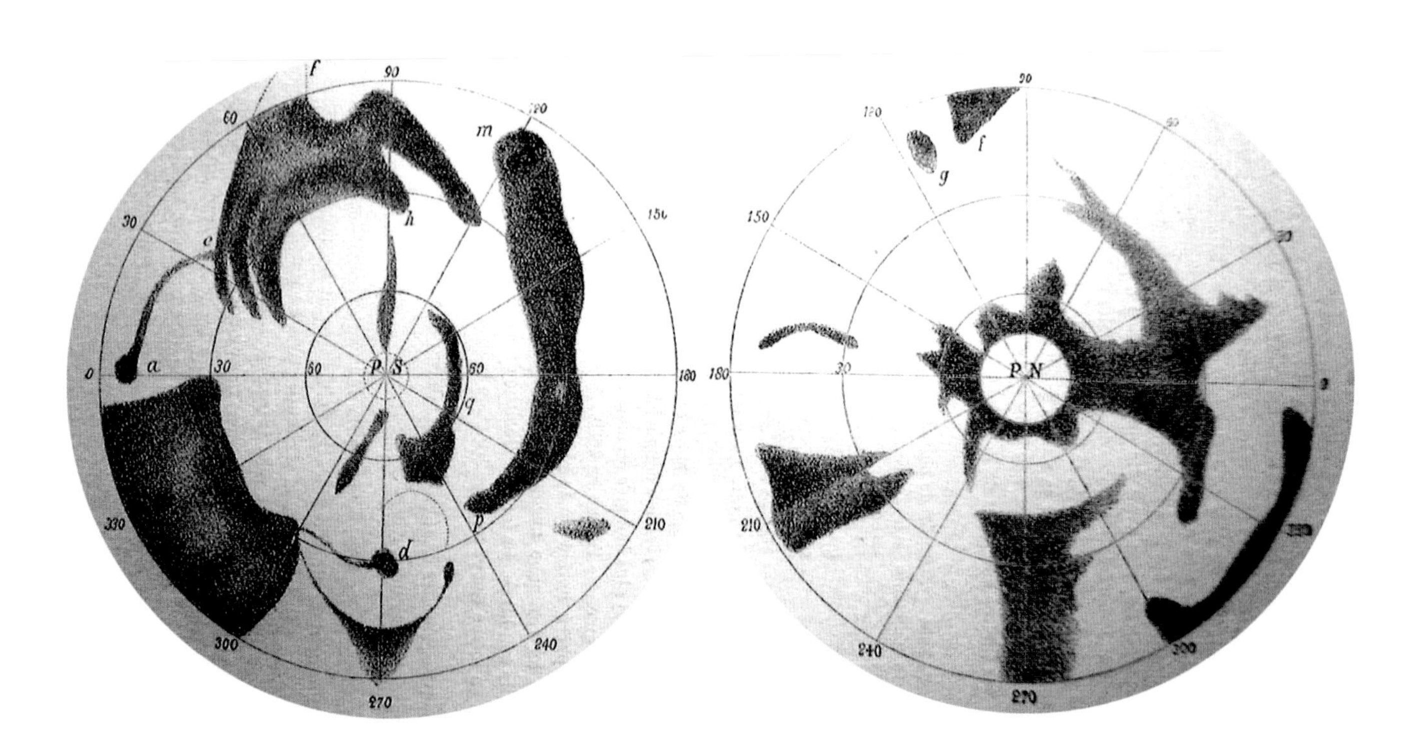

RICHARD PROCTOR, CONTINENTS OF MARS, 1867

The British astronomer Richard Proctor used 27 drawings of Mars produced by William Dawes to construct his map of the planet (see top of facing page). Proctor assumed that the dark areas were sea and the lighter areas continents, with ice caps at each of Mars' poles. He named the features he saw, taking the chance to honour those astronomers who had examined Mars before him, such as Cassini, Herschel, Beer and Dawes. Proctor was part of the move towards considering Mars habitable. His delineation of 'continents' helped to promote that view, and he considered Mars rather pointless if it wasn't inhabited, writing in *Other Worlds Than Ours*: 'Processes are at work out yonder in space which appear utterly useless, a real waste of Nature's energies, unless, like their correlatives on earth, they subserve the wants of organized beings.'

ÉTIENNE TROUVELOT, MARS, 1882

The French artist and astronomer Étienne Trouvelot was astonishingly prolific, producing 7,000 astronomical illustrations from his observations. His beautiful, accurate depictions led to him being invited to work at the Harvard College Observatory and later to use the 26-inch telescope at the U. S. Naval Observatory.

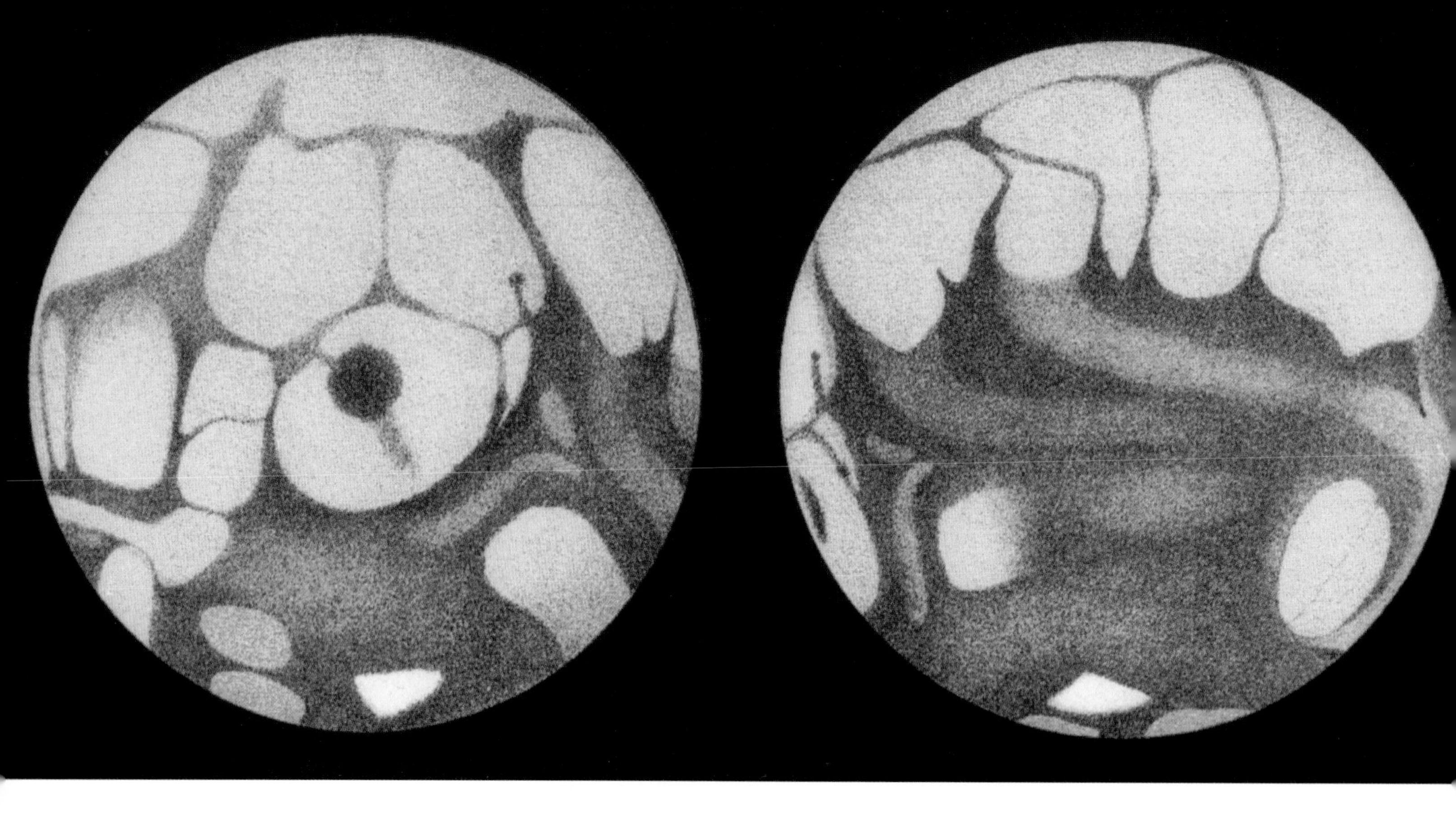

GIOVANNI SCHIAPARELLI, MARTIAN 'CANALI', 1892

The notion that Mars might be inhabited received a great boost from a series of maps produced by Italian astronomer Giovanni Schiaparelli, which apparently showed a network of straight lines. As time progressed, the lines became increasingly prominent in Schiaparelli's maps. His choice of the word 'canali' to describe them caused confusion; while in Italian the word just means 'channels', English readers took it to mean 'canals' – artificially constructed waterways.

Schiaparelli believed them to be naturally occurring features. In 1893, he wrote: 'It is not necessary to suppose them the work of intelligent beings, and, notwithstanding the almost geometric appearance of all of their system, we are now inclined to believe them to be produced by the evolution of the planet, just as on Earth we have the English Channel and the channel of Mozambique.' Others were not so happy to account for them in this way, and argued that if Mars has a canal system it must have (or have had) Martians to make them.

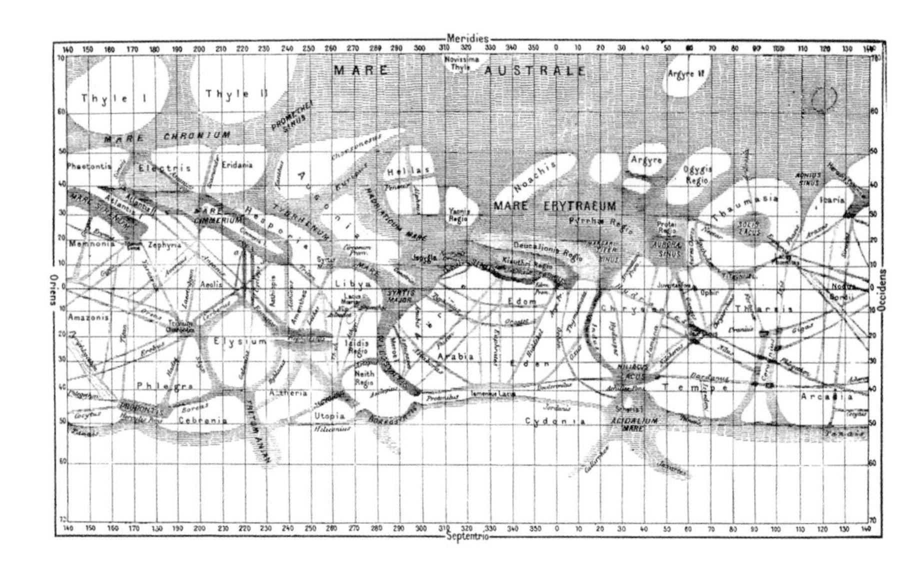

PERCIVAL LOWELL, CANALS ON MARS, 1895

Lowell was an amateur astronomer, greatly excited by the prospect of canal-building Martians. He built an entire observatory in Flagstaff, Arizona, in order to examine Mars and its 'canals', and concocted a fanciful explanation for them. He came up with a theory that the planet was running out of water; in desperation, the Martians had been driven to excavate a complex network of canals so they could tap the polar ice caps and deliver water to areas of their planet in need of irrigation. Astronomers were, in the main, sceptical, but the public loved Lowell's idea. In his book on Mars, he named 184 canals, sixty-four 'oases', and forty regions of the planet.

EARL SLIPHER, MARS FOR MARINER, 1962

The canals were a persistent feature of maps of Mars even long after it had been shown that they were an optical illusion caused by early low-resolution telescopes. This map produced by Earl Slipher, which was used to help plan the Mariner missions to Mars, still shows what look like canals. The Mariner mission proved conclusively that there was no such thing as canals on Mars. Six inset polar maps tilted at varying angles show summer in the northern and southern hemispheres.

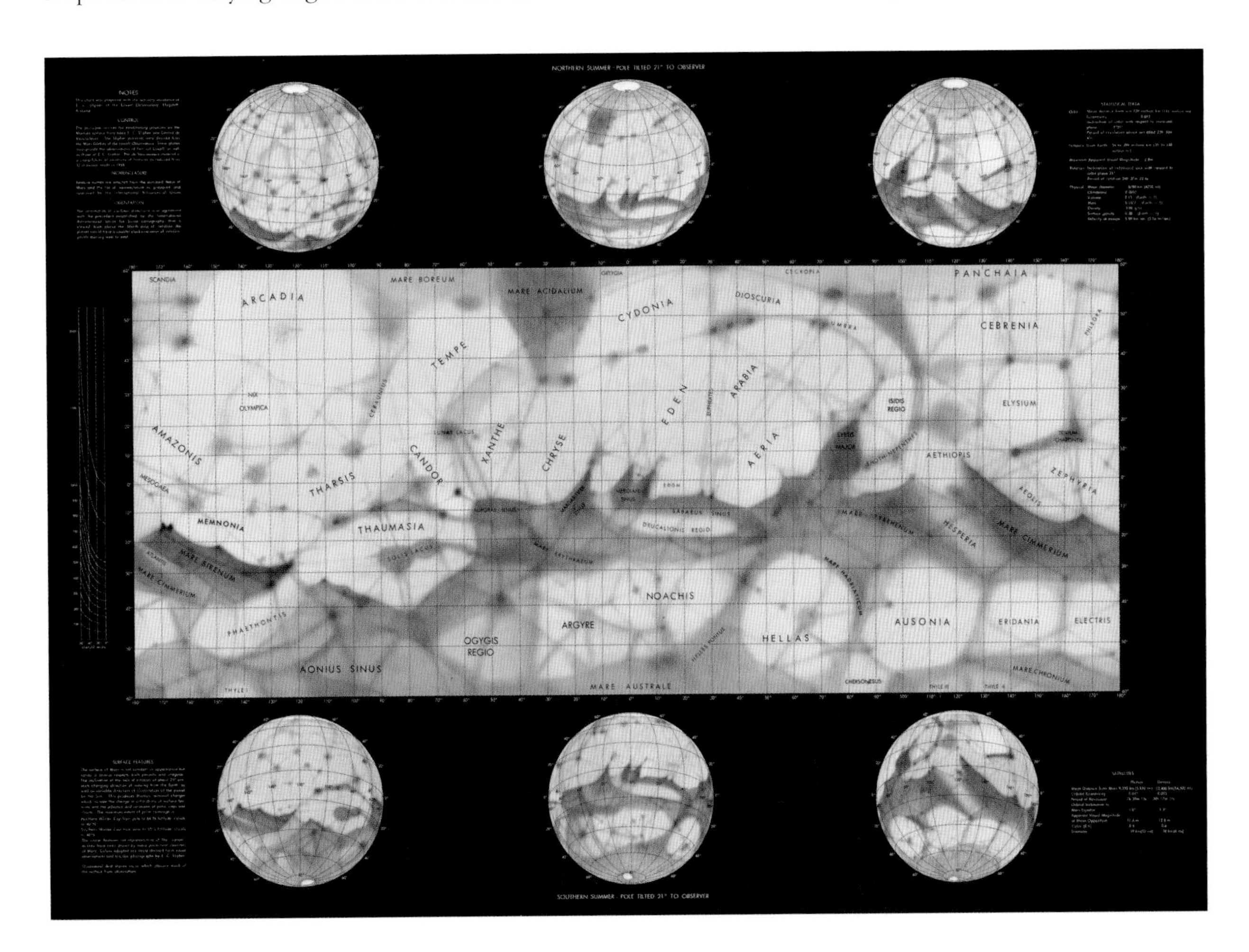

MARINER 9, SURFACE OF MARS, 1971–2

In 1965, NASA's Mariner 4 achieved the first flyby of Mars. The early Mariner missions returned 200 photos of the planet's surface, taken by an orbiter, which were enough to demonstrate that there are no canals. Mariner 9 took 7,000 photographs from which the first detailed map of the surface was constructed. The Mariner missions revealed volcanoes, lava flows, evidence of landslides, polar ice caps and dust storms – but no cities, irrigation canals or desperate Martians. Also visible from the photos was *Olympus Mons* (then called *Olympus Nix*), which at twice the height of Mount Everest is the largest volcano in the solar system.

Mariner 9 was the first spacecraft to orbit another planet, and immediately delivered surprise results. It reported back that the surface was completely hidden by an atmosphere thick with dust, whipped up by the largest storm ever known. Only the four tallest mountain peaks broke through the top of the dust. Imaging had to be delayed for two months while the dust settled. After 349 days in orbit, Mariner had imaged 80 per cent of the planet's surface. As well as the large features, it revealed evidence of erosion by wind and water, fog, weather fronts and ice clouds. Mariner's findings underpinned planning for the later Viking mission to Mars.

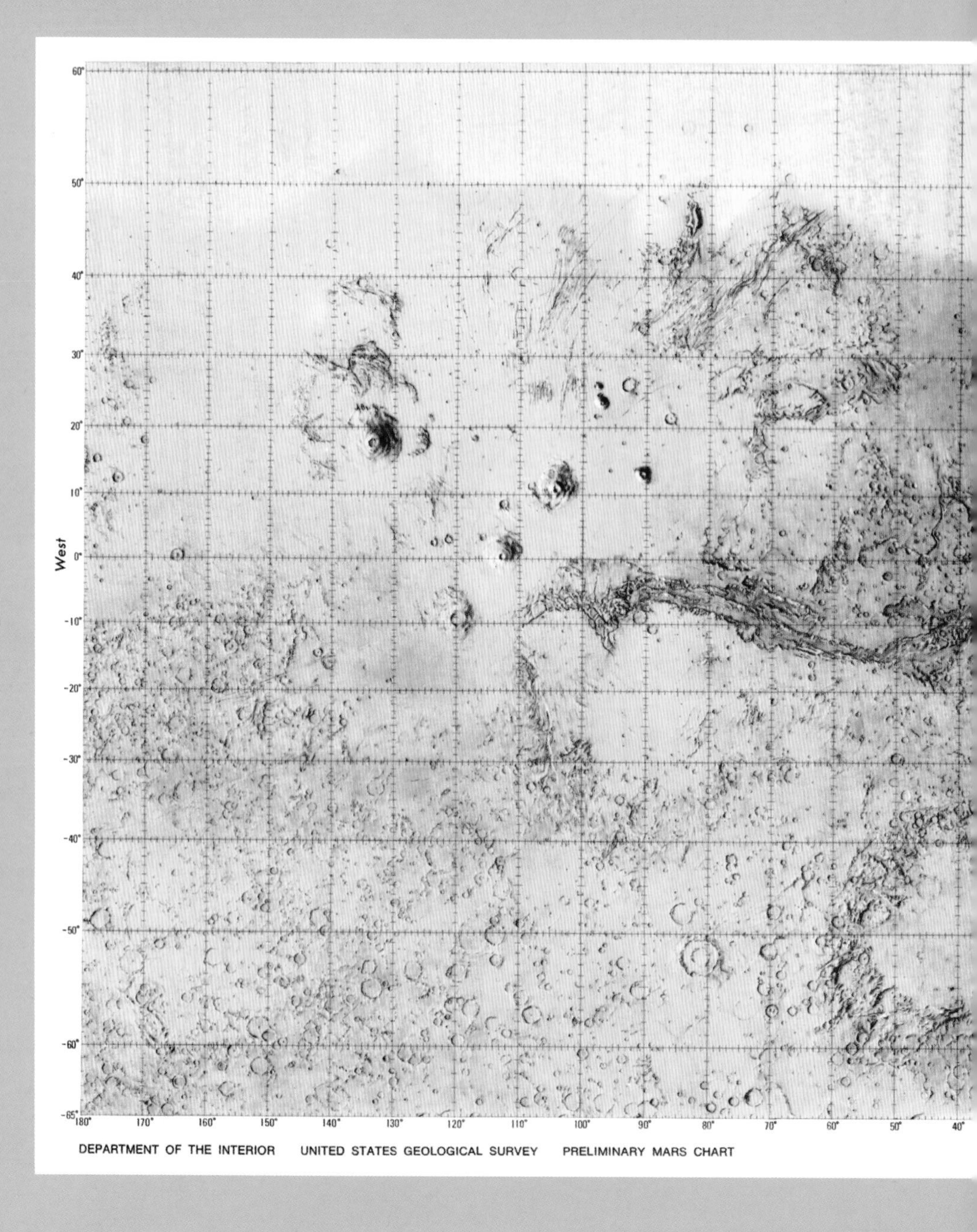

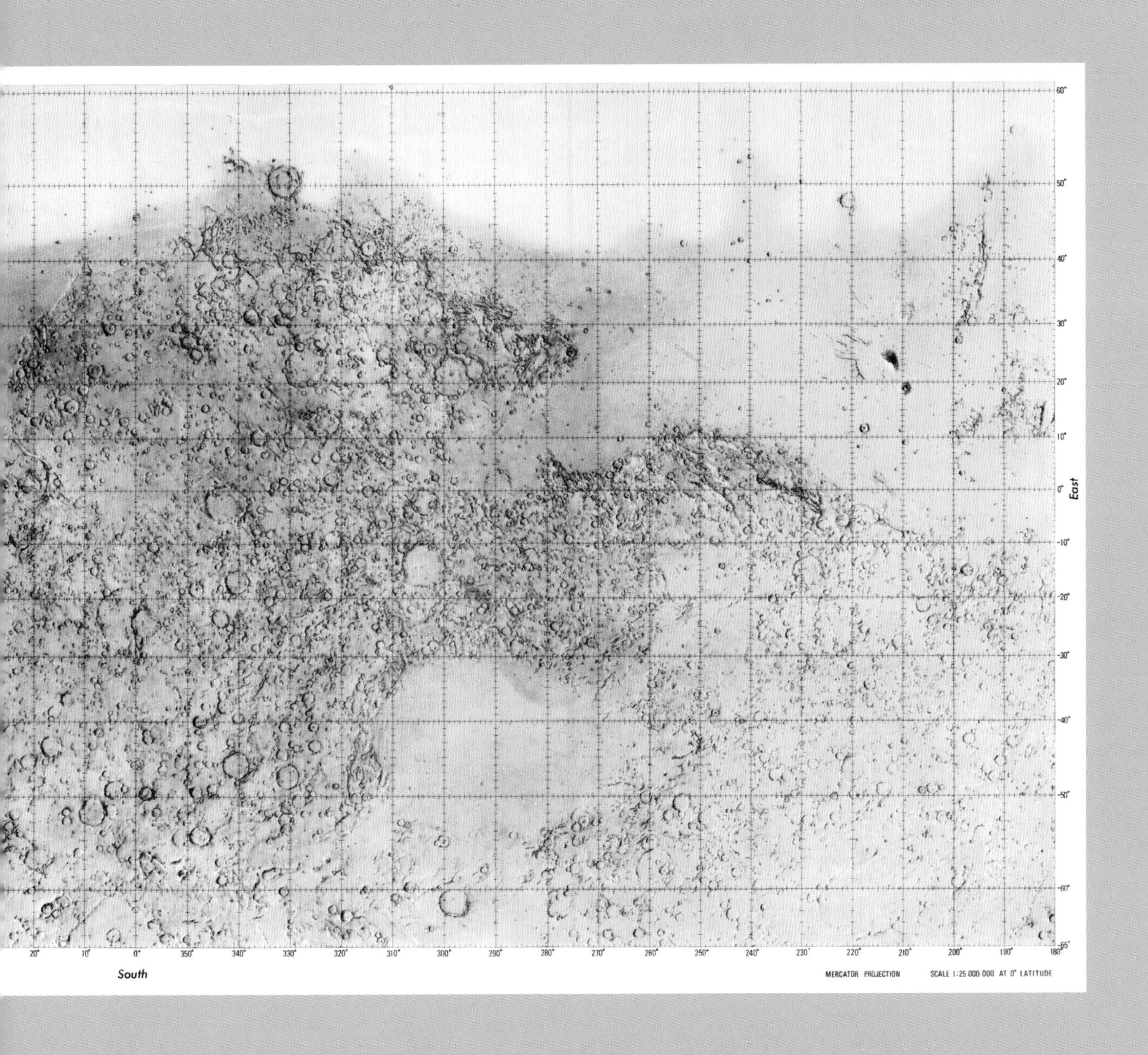

60°
50°
40°
30°
20°
10°
0°
East
-10°
-20°
-30°
-40°
-50°
-60°
-65°
20°
10°
0°
350°
340°
330°
320°
310°
300°
290°
280°
270°
260°
250°
240°
230°
220°
210°
200°
190°
180°
South
MERCATOR PROJECTION
SCALE 1:25 000 000 AT 0° LATITUDE

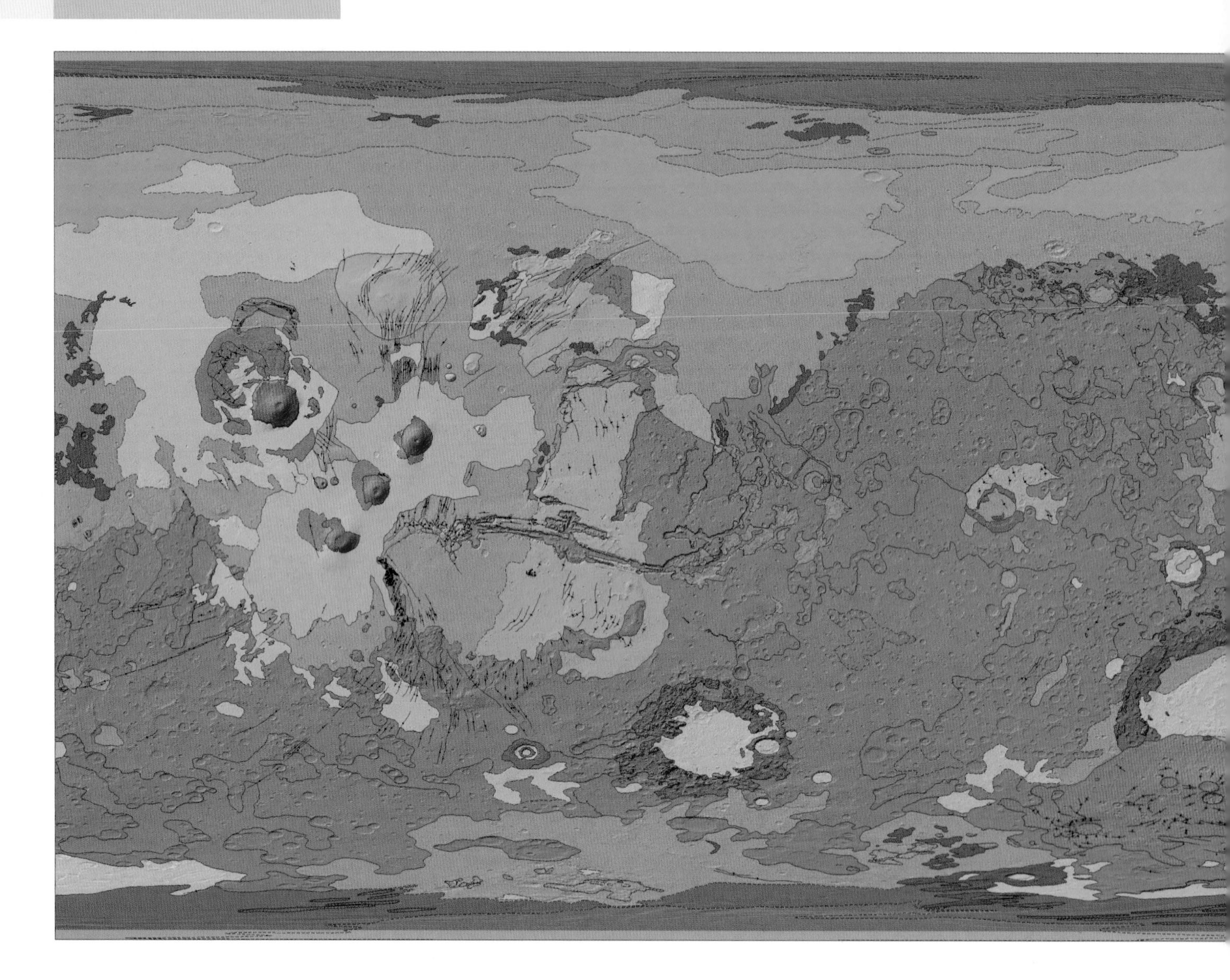

USGS, GEOLOGICAL MAP OF MARS, 1978

The first geological map of any planet other than Earth was produced by the US Geological Survey from the Mariner data. It doesn't show the composition of the planet – there wasn't enough data about the minerals present on Mars for that – but it is colour-coded to show different types of terrain. Blue areas are hilly and cratered, with the dark purple areas representing mountains. Light yellow areas are smooth plains and darker yellow indicates cratered plains. Volcanic material is shown in pink.

The abundance of volcanoes and lava flows shows that Mars has had a geologically active past (and could still be geologically active), while the number of craters indicates that its surface is very old. A frequently renewed surface would lose evidence of previous bombardment.

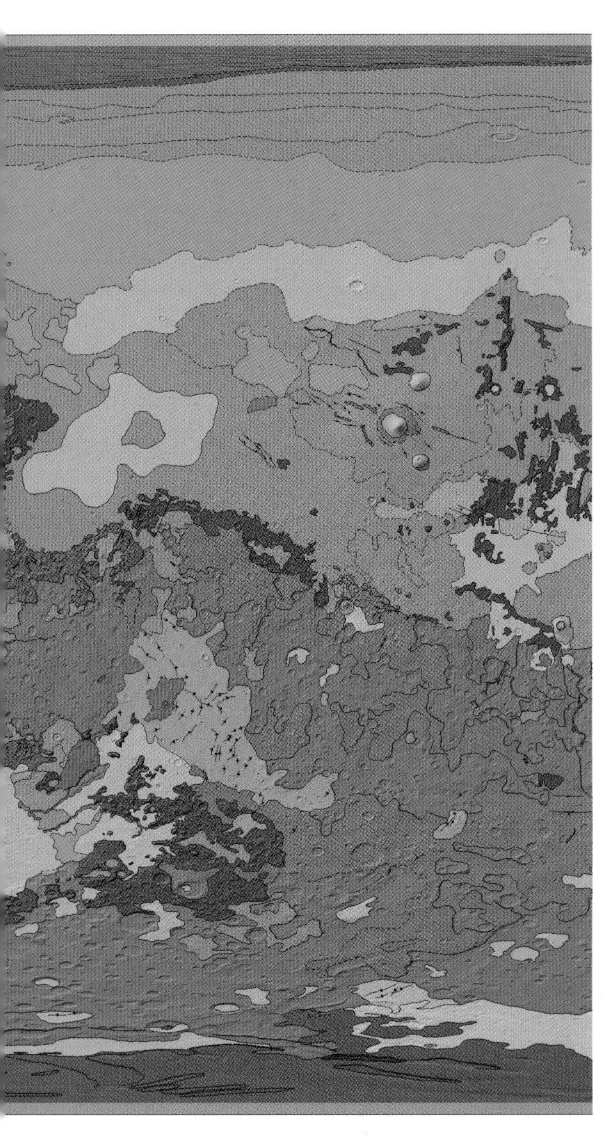

USGS, GEOLOGY OF MARS, 2014

USGS has taken data from the Mars Global Surveyor, Mars Odyssey (2001), Mars Express (2003) and the Mars Reconnaissance Orbiter (2006) and combined it to make the most detailed geological map of Mars to date. The brown area is the oldest area of surface, and was laid down 3.7–4.1 billion years ago. Evidence from the landers suggests that Mars has remained volcanically active until very recently or up to the present. There is also evidence of quakes, glaciers and water erosion.

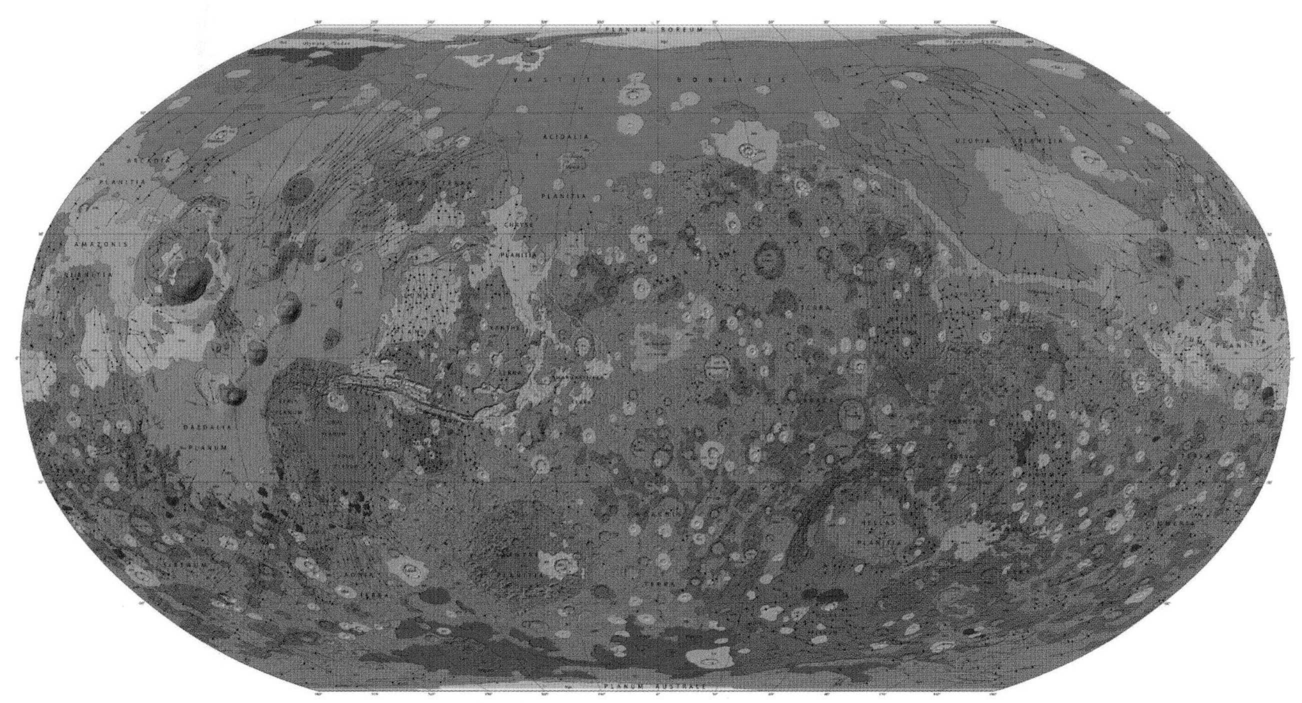

NASA, TOPOGRAPHY, GRAVITY, CRUSTAL THICKNESS AND SURFACE TERRAIN COMPARED, 2016

These five maps together capture the geological character of Mars. They show how the surface terrain (in natural colour, map 5) relates to topography (map 1), the thickness of the Martian crust (map 3) and the gravity profile of the planet (maps 2 and 4).

The data was collected by analyzing the flight paths of three spacecraft in orbit around Mars: the Mars Global Surveyor, Mars Odyssey and Mars Reconnaissance Orbiter. As a spacecraft orbits Mars, tiny variations in the planet's gravity cause slight changes in the craft's altitude and speed. These can be monitored to produce a map that reveals areas with higher or lower gravity. The variations in gravity are caused by the bumpy,

1

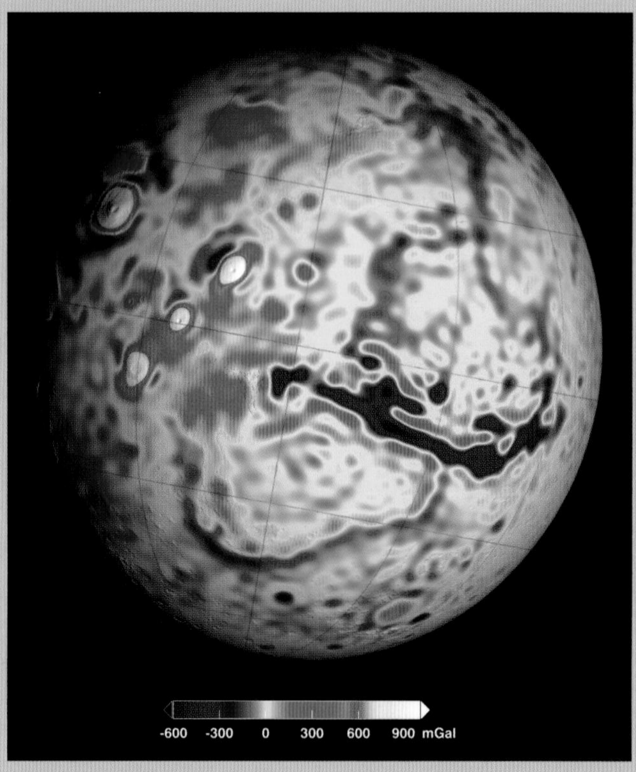

2

3

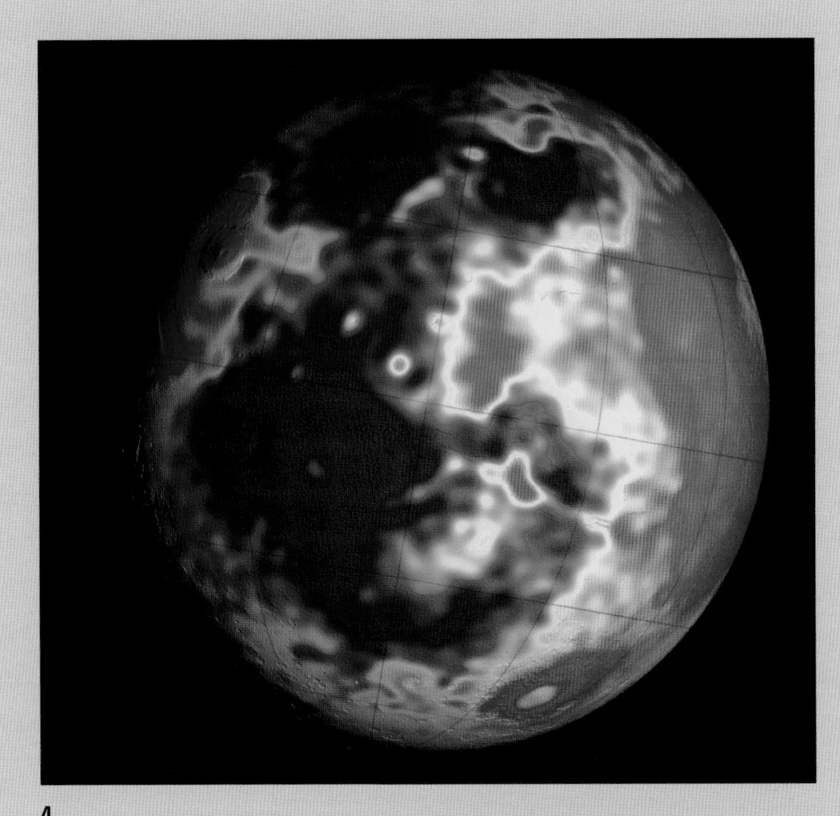

4

uneven surface of the planet and the lumpiness of its interior. Higher gravity can be caused by higher altitude on the surface (so there is more land between the orbiter and the planet's centre of gravity) or by denser material within the planet.

In the gravity map (2), the long blue streak is a canyon, *Valles Marineris*, an area with low gravity. The red and white areas have the highest level of gravity, including the three peaks of the Tharsis volcanoes, *Arsia Mons*, *Pavonis Mons* and (northernmost) *Ascraeus Mons*.

The free-air gravity map (2) directly records the differences in gravity. The Bouguer gravity map (4) subtracts the bumpiness of the surface from the effect and shows the lumpiness of the planet's interior, with dense areas below the surface increasing the gravity. The map of crustal thickness (3) was calculated from the Bouguer map, improving previous estimates of the thickness.

The measurements also confirmed that Mars has a molten outer core.

The gravity maps will help future missions plan their entry into Mars orbit and approach to the planet accurately.

5

MARS EXPRESS, *MANGALA FOSSAE*, ORTHO-IMAGE, 2008

The *Mangala Fossae* is a system of outflow channels apparently created by catastrophic flooding up to 3.7 billion years ago. It is thought to have been created when a system of dykes (lava channels) from the nearby Tharsis volcanoes melted a huge frozen reservoir beneath the surface. On at least two occasions, huge volumes of water cascaded down the *fossae* onto the Martian plains. In this image, the higher land (white/grey) is lava deposits. The lower land (blue) is a flat, depressed plain

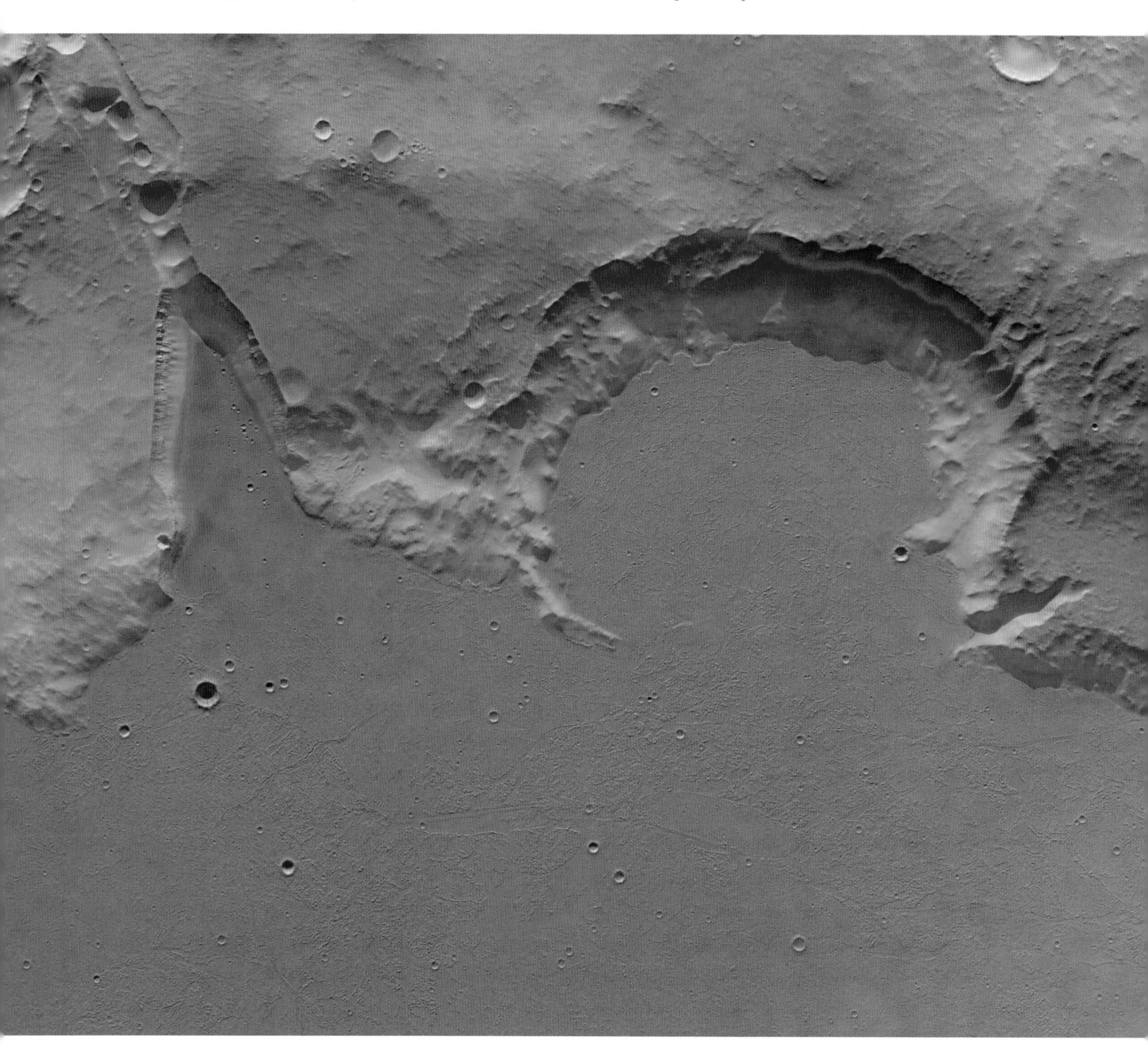

where the water once collected; its even, uncratered surface shows that it is relatively young. The cliff edges are up to 100 m (328 ft) high in places.

An ortho-image is one that has been corrected to remove the distortion of an aerial image, so that true measurements can be taken from it. The photographs from which this image was created were taken by ESA's Mars Express, using its High Resolution Stereo Camera.

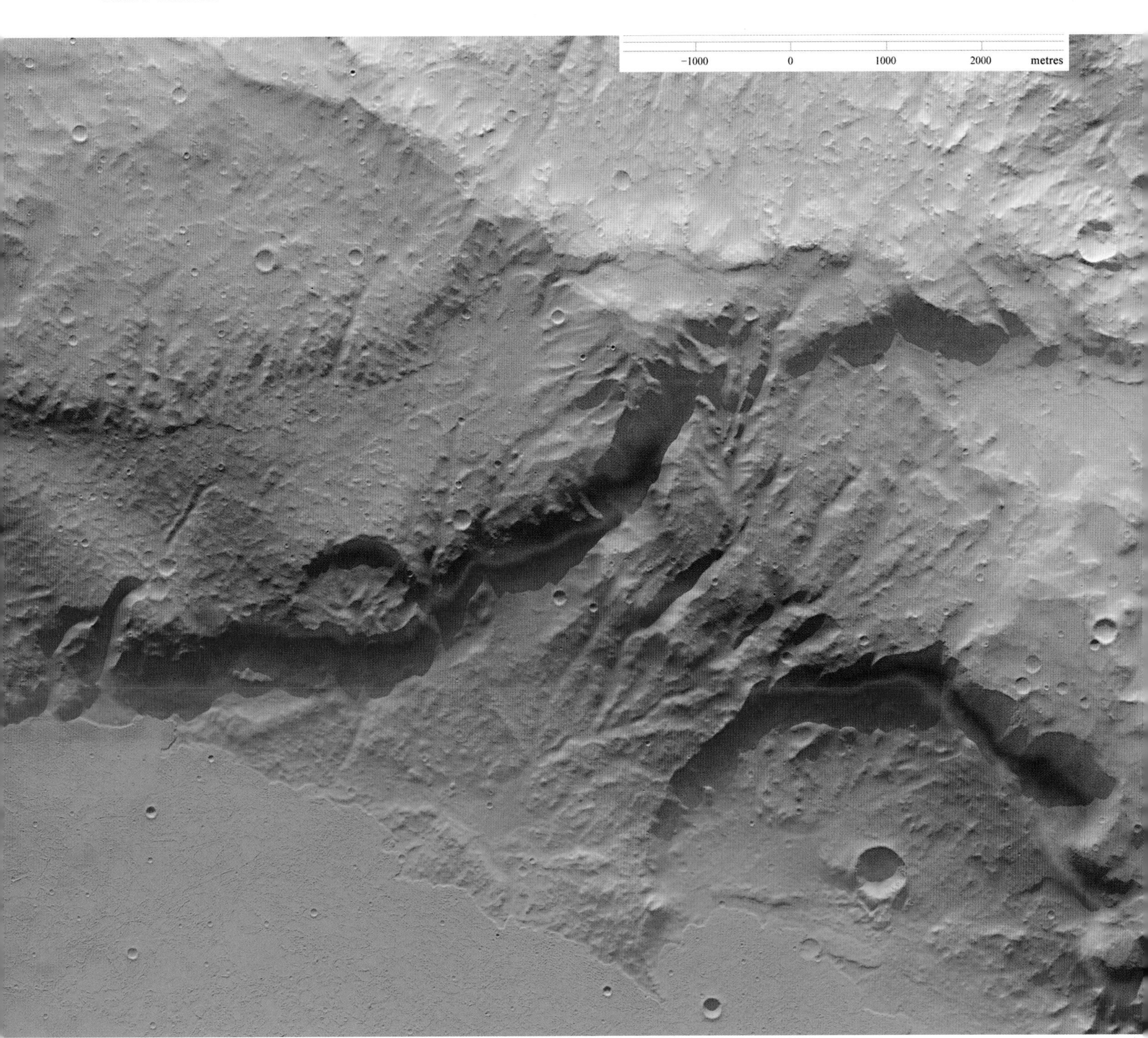

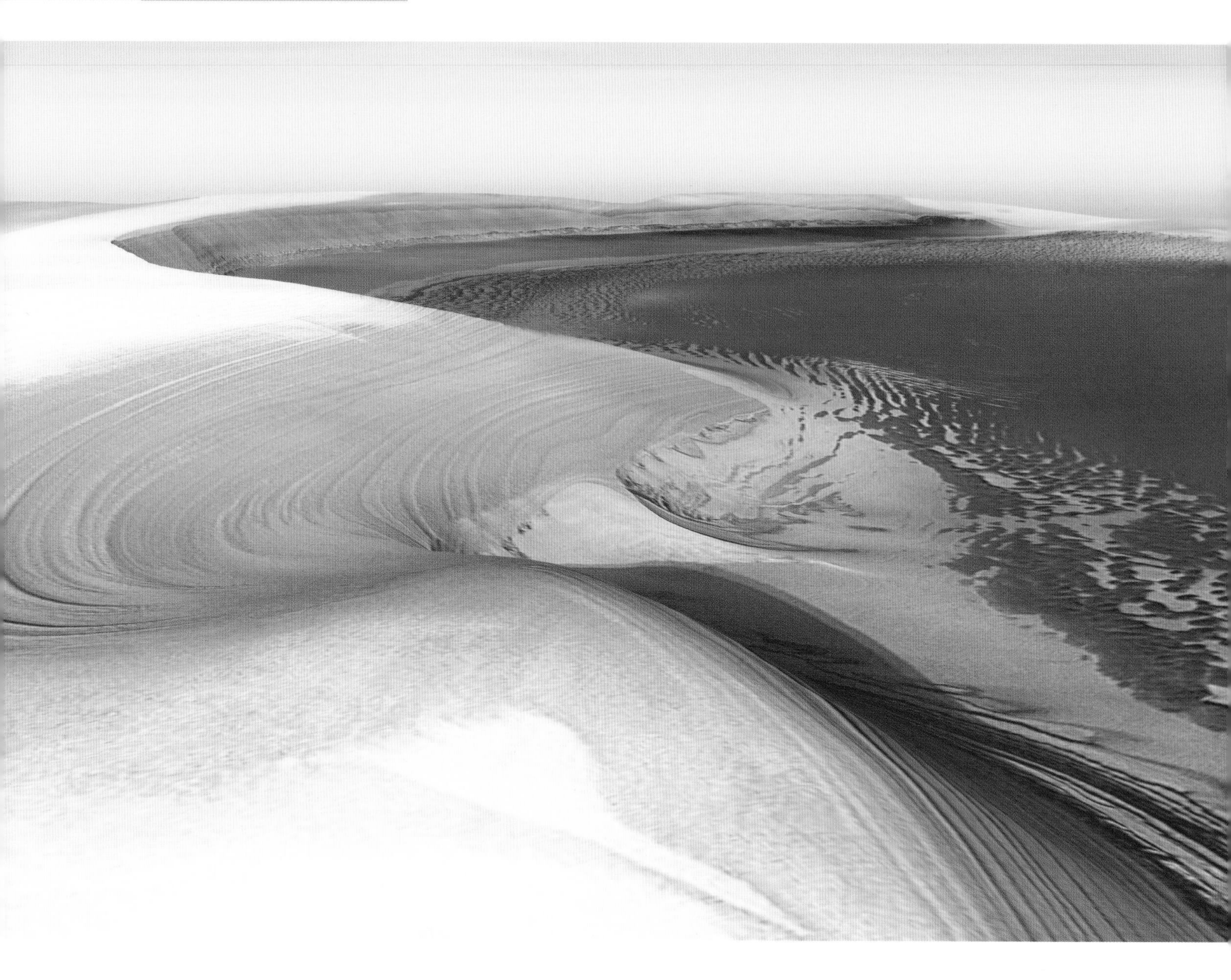

HiRISE, MARS IN CLOSE-UP, 2002–5

The High-Resolution Imaging Science Experiment on NASA's Mars Reconnaissance Orbiter took 13,000 detailed photographs covering about one per cent of the surface of Mars. The images are very high resolution, with one pixel corresponding to an area about 25–60 cm (10–24 in) on the ground (depending on the height of the orbiter above the surface).

Mars has one of the most rugged surfaces in the solar system. The main image shows part of the *Chasma Boreale*, a canyon that cuts through 570 km (350 miles) of the northern polar ice cap. The walls of the canyon are up to 1,400 m (4,600 ft) tall. The image shows a mix of sand and ice.

The tilt of the axis of Mars varies over hundreds of thousands of years, leading to cyclical climate change. The ice at the poles has accumulated and been eroded over billions of years, building during cold spells and being eaten away in warmer spells. Sand blown in during warmer times has frozen into the ice. It is exposed and blown into the canyon as the ice melts, forming dunes. As the ice is currently melting, a relic sand landscape beneath the polar ice will gradually be revealed.

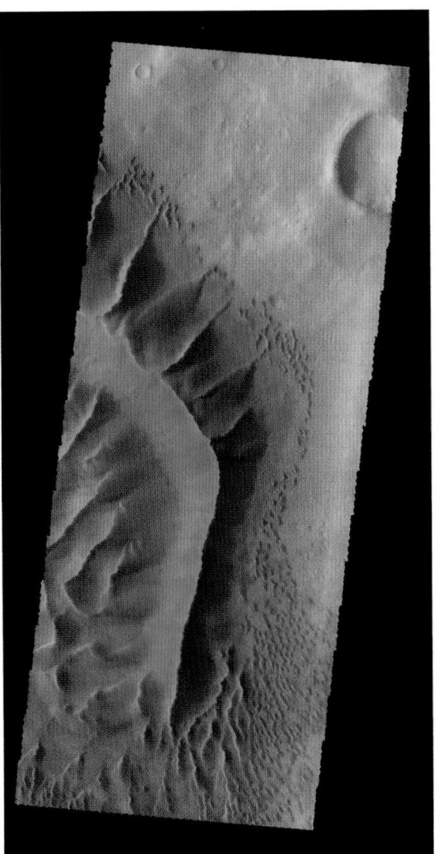

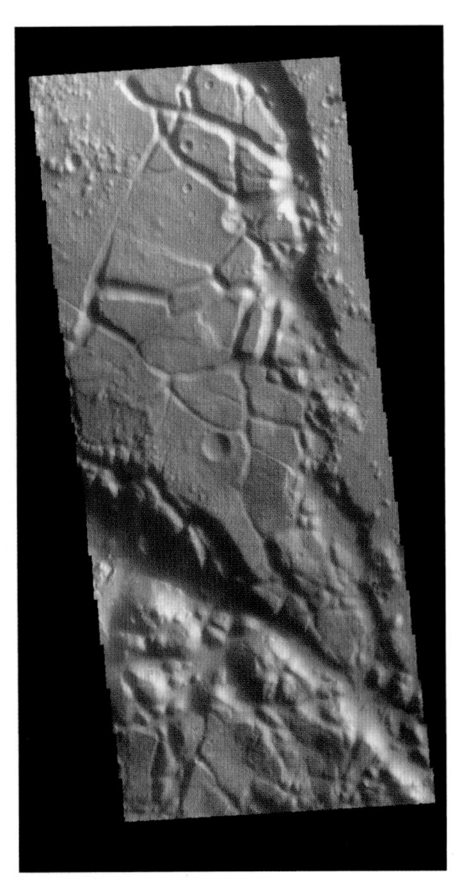

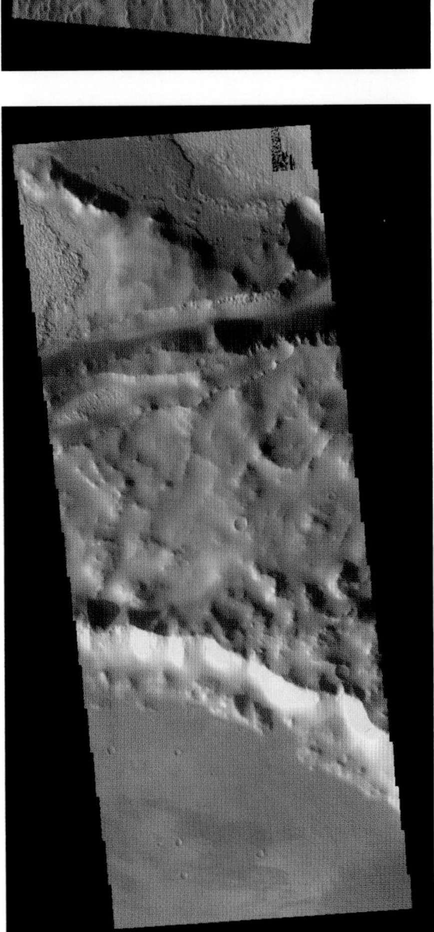

Views from a greater altitude show the shape of the canyon. The image at the top left shows most of its extent. The canyon widens to 120 km (75 miles) at its mouth.

MOONS OF MARS

Mars has two very small moons, Phobos and Deimos. Phobos is about 22 km (14 miles) across and Deimos just 12 km (7.5 miles) across.

Both moons are much closer to Mars than our Moon is to Earth. Phobos orbits 6,000 km (3,700 miles) and Deimos orbits 20,000 km (12,500 miles) above the planet. Compared with the Moon, which is 383,000 km (238,000 miles) from Earth, they are extremely close to Mars. While it took the Apollo astronauts three days to travel to the Moon, it would take an equivalent craft about an hour to get from Mars to Phobos. It takes Deimos 30 hours to orbit Mars, but it only takes Phobos around eight hours.

CURIOSITY, PHOBOS, 2013

The image of Phobos (right) was taken from NASA's Curiosity rover. The moon's prominent feature is the Stickney crater, lower right in this image. Phobos is a doomed moon; orbiting so close to Mars that it will be drawn closer and eventually (in around 50–100 million years) either crash into the planet's surface or be broken up.

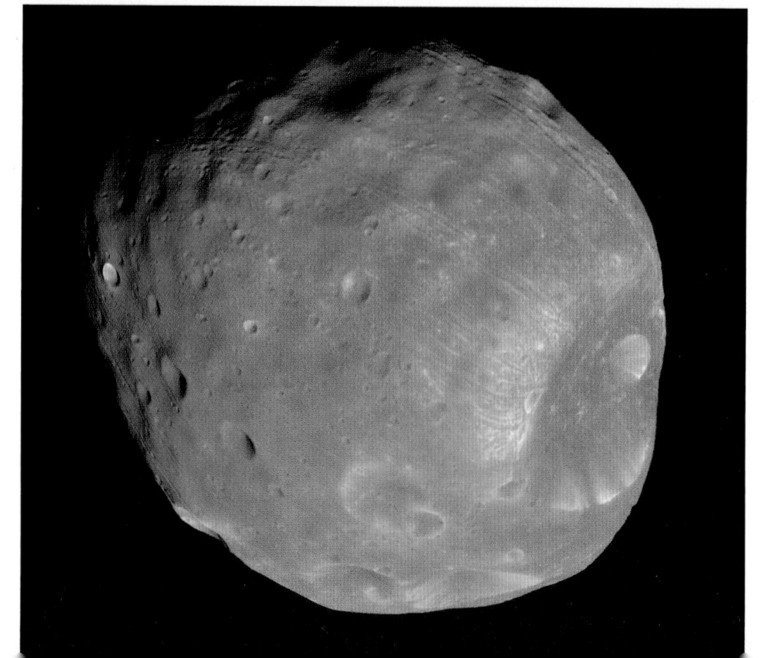

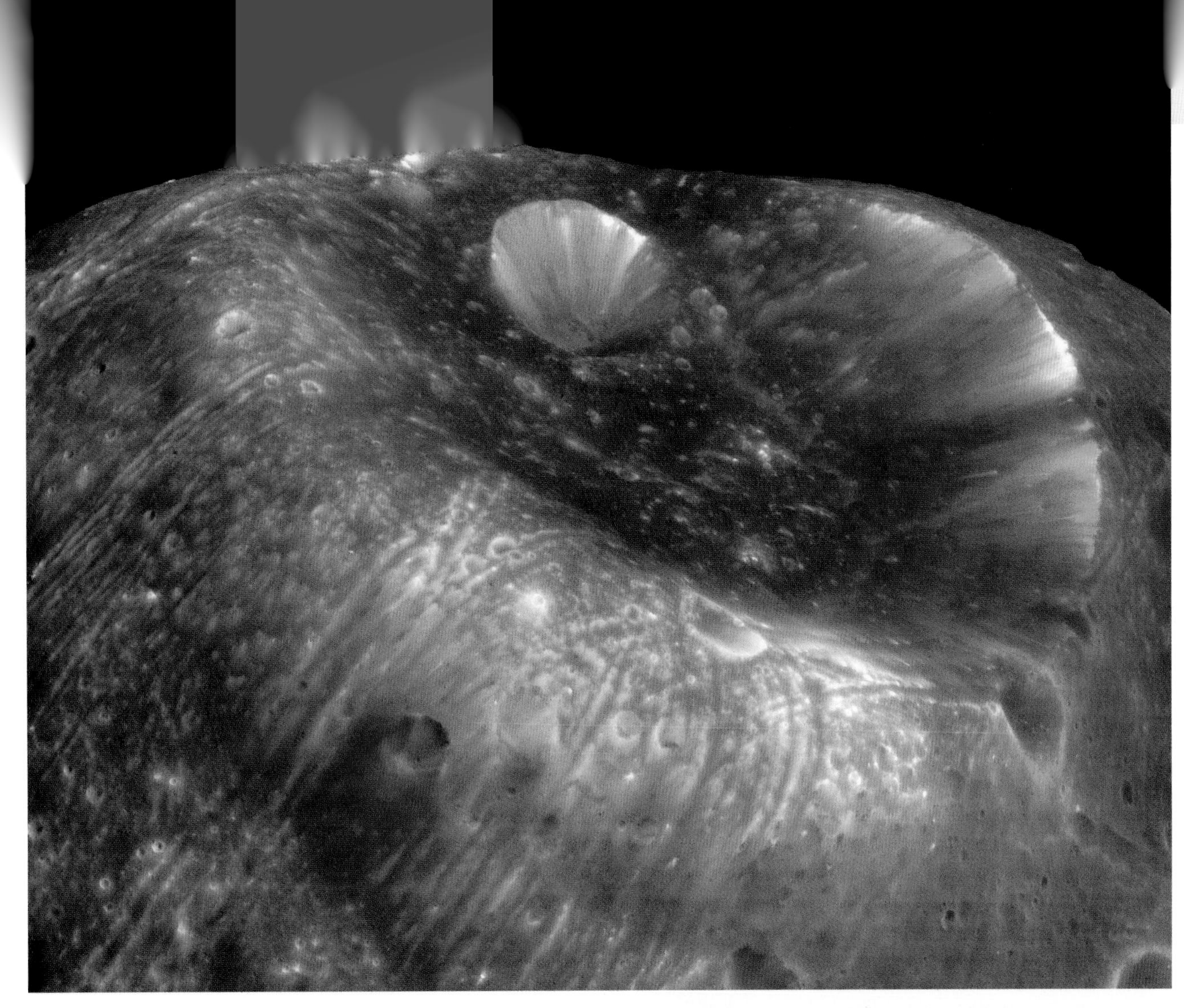

Above: *A false-colour image of the Stickney crater on Phobos, photographed by the HiRISE camera onboard the Mars Reconnaissance Orbiter in 2008.*

MARS EXPRESS, DEIMOS AND SATURN, 2018

This image of Deimos was taken in 2018 by the Mars Express' High Resolution Stereo Camera. Deimos has only two named features, the craters Voltaire and Swift, after two famous authors who speculated that Mars had two moons (before the discovery of Deimos in 1877). The surface of Deimos is smoother than that of Phobos because many of its craters are filled with regolith (loose material). This photograph was taken looking away from the Sun towards the outer solar system. Saturn, identifiable by its rings, is just visible in the background.

The large, metal-rich asteroid Lutetia (top) is 120 km (74.5 miles) long and heavily cratered. It was photographed by the ESA's Rosetta probe in 2010. The smaller asteroid here is Gaspra, photographed by the Galileo craft in 1991 on the first fly-by of an asteroid.

ESA, THE ASTEROID BELT

The Asteroid Belt is a band of rocky debris left over from the formation of the solar system. Astronomers of the early 19th century searched for a planet between Mars and Jupiter (the distances between the known planets suggested there should be one). Instead, the asteroids of the Asteroid Belt were found in its place. At about 1 AU (150 million km/93 million miles) across, the Asteroid Belt is the source of many of the meteoroids which fall to Earth.

Some asteroids are solid lumps of rock, others are collections of rubble held together by their gravity. A small proportion are rich in metals, particularly iron and nickel and some are made of a mix of rock and metal. Relatively few asteroids have been mapped or imaged in detail. They vary in size from the dwarf planet Ceres, 946 km (588 miles) across, to objects the size of pebbles. Nearly a million have been identified since Giuseppe Piazzi found the first, Ceres, in 1801.

GALILEO, IDA AND DACTYL, 1993

This photograph of Ida and its moon Dactyl was taken by the robotic spacecraft Galileo en route to Jupiter. Dactyl was the first moon of an asteroid ever discovered. Ida, the potato-shaped asteroid, is 56 km (35 miles) long and around 23 km (14 miles) across, while its egg-shaped moon is just 1.6 km (1 mile) across at its largest. Ida's rotation period (day length) is 4 hours 38 minutes.
Ida is a stony or stony/iron asteroid believed to have been created when a much larger asteroid (named Koronis) was smashed apart in a collision. Ida's heavily cratered surface suggests it is at least a billion years old. Dactyl seems to be younger and of a slightly different composition. Its regular shape suggests it has agglomerated from a collection of smaller particles; one theory is that it was created from a debris cloud that formed around Ida after a meteor impact.

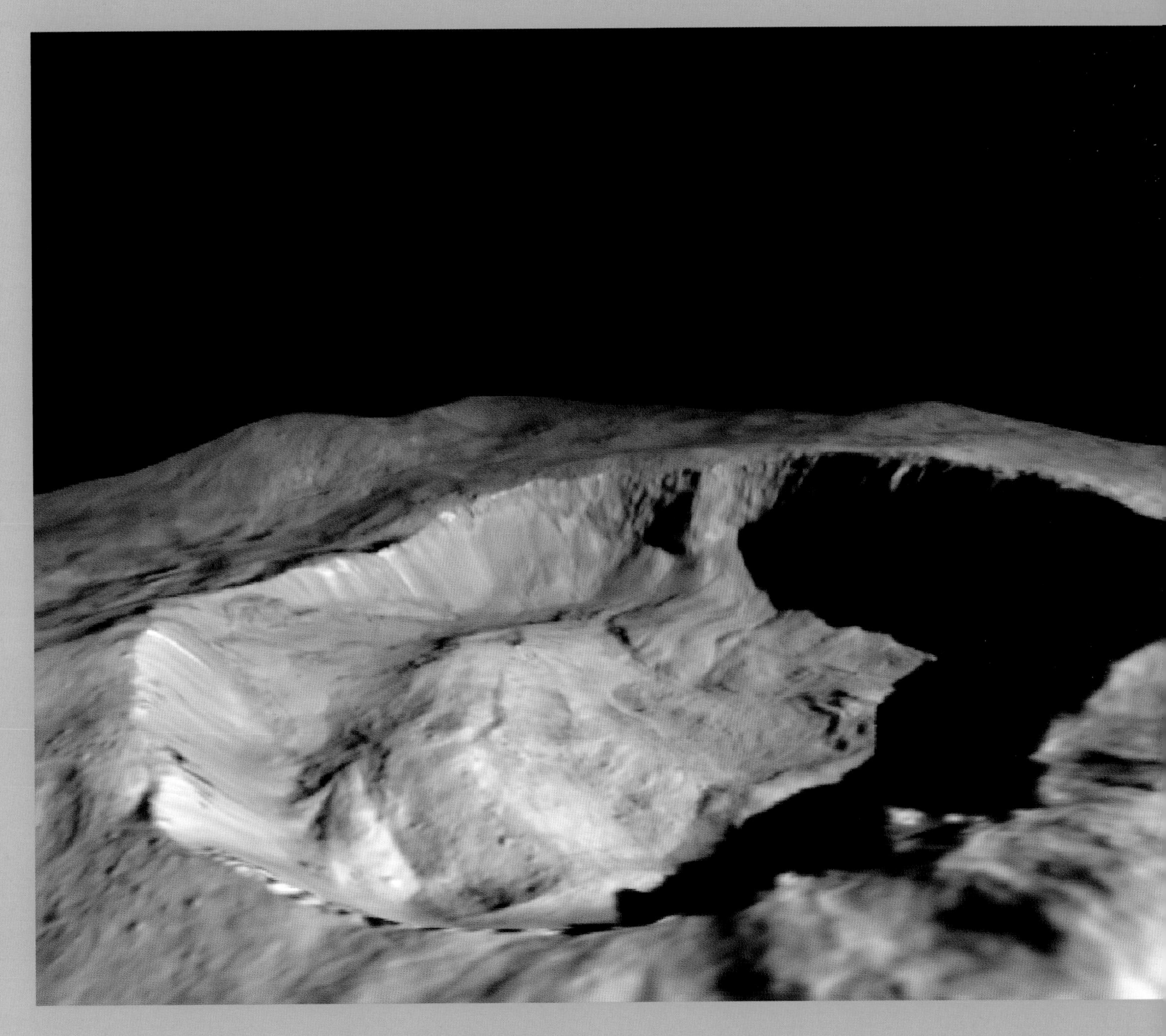

DAWN, CERES, 2015

The two largest objects in the Asteroid Belt, Ceres and Vesta, were the target of NASA's Dawn mission – the first to orbit a dwarf planet and to orbit two objects. Ceres has a water-rich crust and could be as much as 25 per cent water – in which case it would contain more water than Earth. Dawn confirmed that Ceres has a coat of ice wrapped around a rocky core.

Ceres is the only dwarf planet in the Asteroid Belt; the others lie beyond Neptune. The Juling crater, shown here, is 2.5 km (1.6 miles) deep. Ice collects on the northern wall, which is in permanent shadow.

DAWN, VESTA, 2011–12

Vesta, the second largest asteroid at 530 km (329 miles) across, formed just 1–2 million years after the start of the solar system. The findings of the Dawn mission confirmed that it is differentiated into layers (a crust, mantle and core), unlike most asteroids, but similar to planets such as Earth. This was possible because it formed so early, when some radioactive elements in its make-up were still sufficiently active to melt the forming asteroid and allow the heavier material to migrate to the middle.

Vesta has dark and light patches on its surface. The light material is thought to be native rock and the dark material may have been deposited by the 300 asteroids that have collided with it in the last 3.5 billion years. As a very early body, it is potentially a valuable object of study for astronomers working on the formation of the solar system. Vesta is the brightest asteroid and sometimes visible to the naked eye.

Above: *This colour-coded topographical map of Vesta is from the Dawn mission.*

The troughs at the top of Vesta in the image on the facing page are canyons. They are possibly the result of fractures from a large impact that created the massive Rheasilvia crater. At 500 km (310 miles) across, this crater is 90 per cent the diameter of Vesta itself. Another trough, the *Divalia Fossa*, is larger than the Grand Canyon. The mountain near the south pole of Vesta, at the bottom of this image, is twice as tall as Mount Everest.

MORE SPACE THAN ROCK

The Asteroid Belt (shown above in its position in the solar system between the orbits of Mars and Jupiter) holds 200 asteroids larger than 100 km (60 miles) across, and 1–2 million larger than 1 km (0.6 miles) across. They are widely spaced, though. If all the contents of the Asteroid Belt were combined, they would make a planet smaller than the Moon.

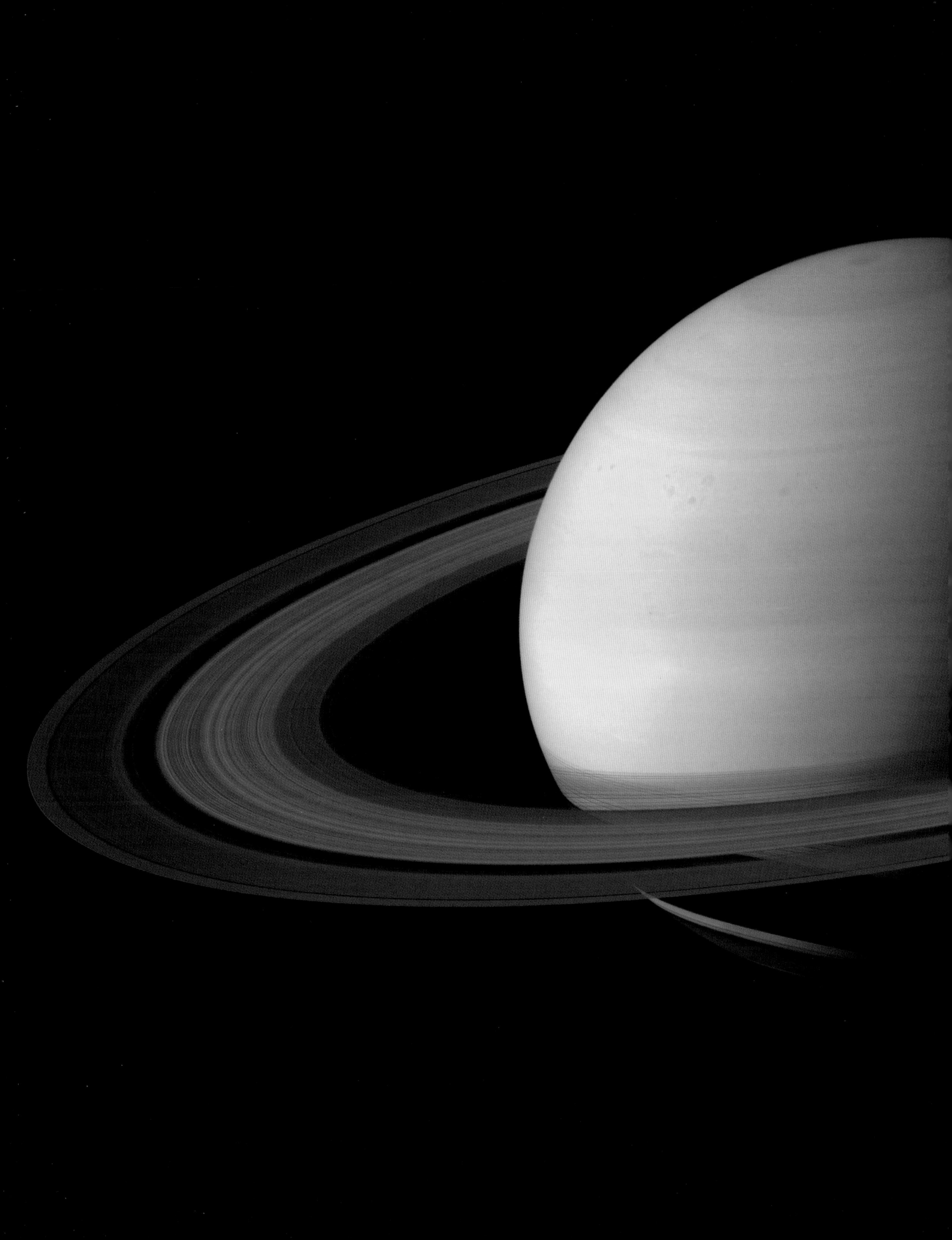

GAS GIANTS

WORLDS OF AIRY NOTHING

The gas giants Jupiter and Saturn are separated from the rocky planets by the Asteroid Belt and a gap of 588 million km (365 million miles), nearly four times the distance from the Earth to the Sun. To many people they are a fascinating enigma – vast balls of gas that don't disperse as gases do on Earth but cling together, wracked by whirling storms as they hurtle through space.

Left: *The gas giant Saturn, photographed by the Cassini spacecraft in 2004. This image is a mosaic made up of 105 separate photos of the planet and its signature rings. The Maori name for Saturn,* Parearau, *means 'surrounded by a headband' – a tantalizing enigma since the rings are visible only through a telescope.*

GAS THROUGH AND THROUGH?

Unlike the rocky planets, the gas giants have no distinct solid surface but that doesn't mean they are gaseous all the way down. For a long time, gas giants were believed to have a small solid core around which the planet originally accreted. The latest data from the Juno mission to Jupiter, however, suggests there might instead be a larger 'fuzzy' core that is at least partially dissolved.

The gas of a gas giant becomes progressively denser, possibly existing in a liquid-like state at some point. It eventually reaches such a density that solid objects will not sink through it, a point described as a 'nominal surface'. This surface differs from that of a rocky planet because there is no clear and consistent boundary between gas and solid. On Earth or Mars, the boundary between surface and atmosphere is indisputable – marked by a change of material (air to rock or water). On Saturn, the hypothetical point at which a human will stop sinking is not the same as the point at which, say, a flower or a spider will stop sinking. When discussing the 'surface' of a gas giant, scientists nominate the point at which the atmospheric pressure is the same as that at Earth's surface (1 bar).

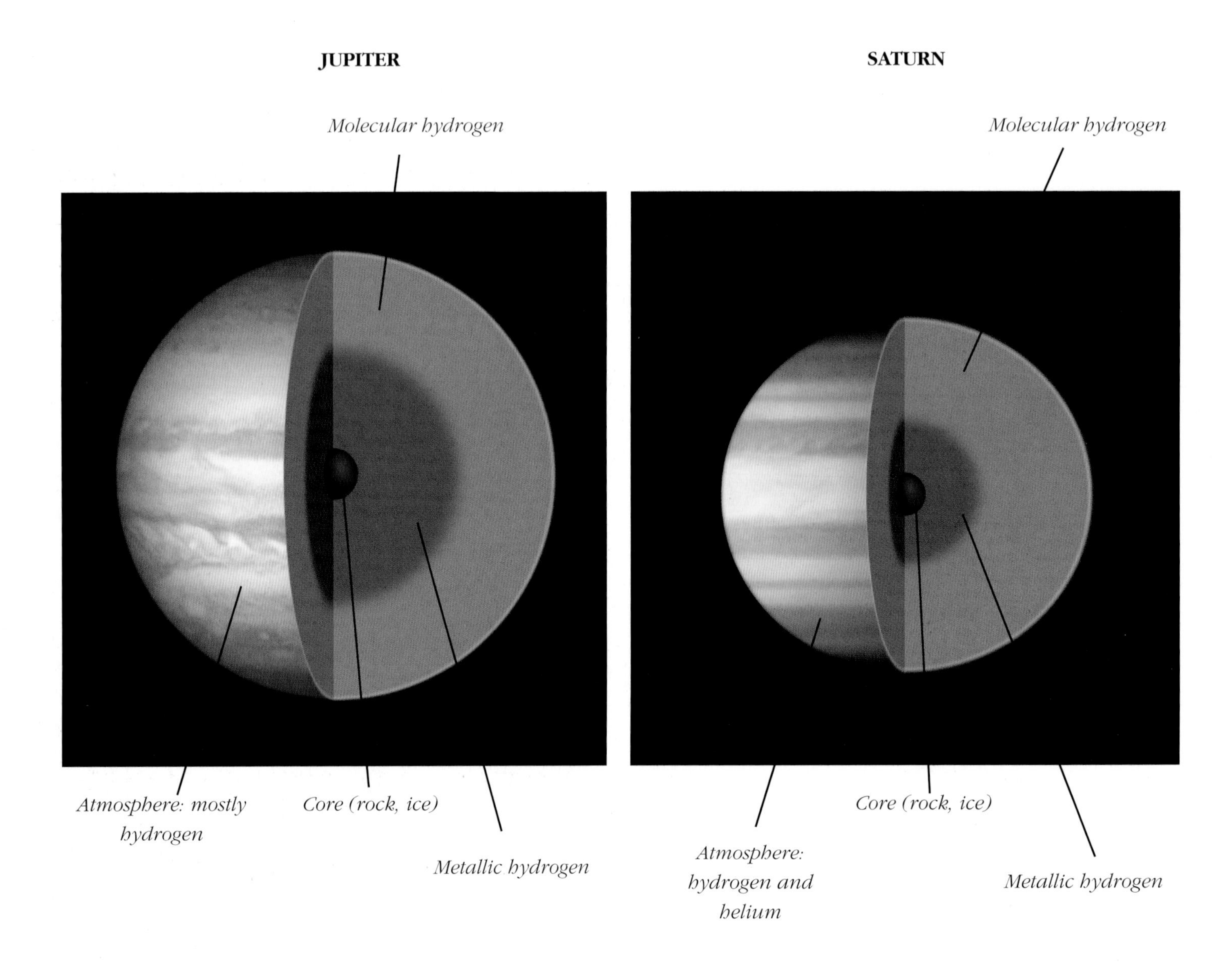

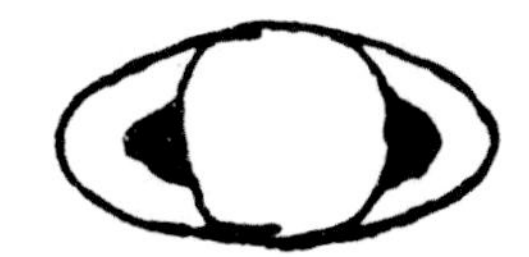

Left: *In 1610, Galileo thought he saw two moons flanking Saturn (third image). In 1616 he drew something now recognizable as a depiction of Saturn's rings (first and second images), but could not work out what he was seeing.*

The gas giants are made predominantly of hydrogen with some helium; they have an atmosphere of swirling clouds driven by storms. Yet each has a surprisingly large coterie of rocky moons. While the planets themselves offer a unique problem to cartographers (how can we map something made of moving gas?), the moons are easier to deal with.

GLIMPSES FROM THE PAST

Like the rocky planets, the gas giants have been known since prehistoric times. Their size and luminosity mean they are visible to the naked eye despite their great distance from Earth. There was nothing to distinguish them from the rocky planets, however. Even with the advent of the telescope, the difference between the two groups was not immediately apparent.

Galileo first saw and drew both Jupiter and Saturn. He could make out something strange near Saturn, which he didn't identify as the planet's rings. And he saw four moons around Jupiter, which had a considerable impact on the contemporary view of the solar system. The moons of Jupiter offered additional support for the Copernican model of the solar system and the not-so-special status of Earth. Most of the moons of the gas planets were too small to be seen even with a telescope. New moons are still appearing today.

The Cassini orbiter launch at Cape Canaveral in October 1997. The spacecraft arrived in the Saturn system on 30 June 2004.

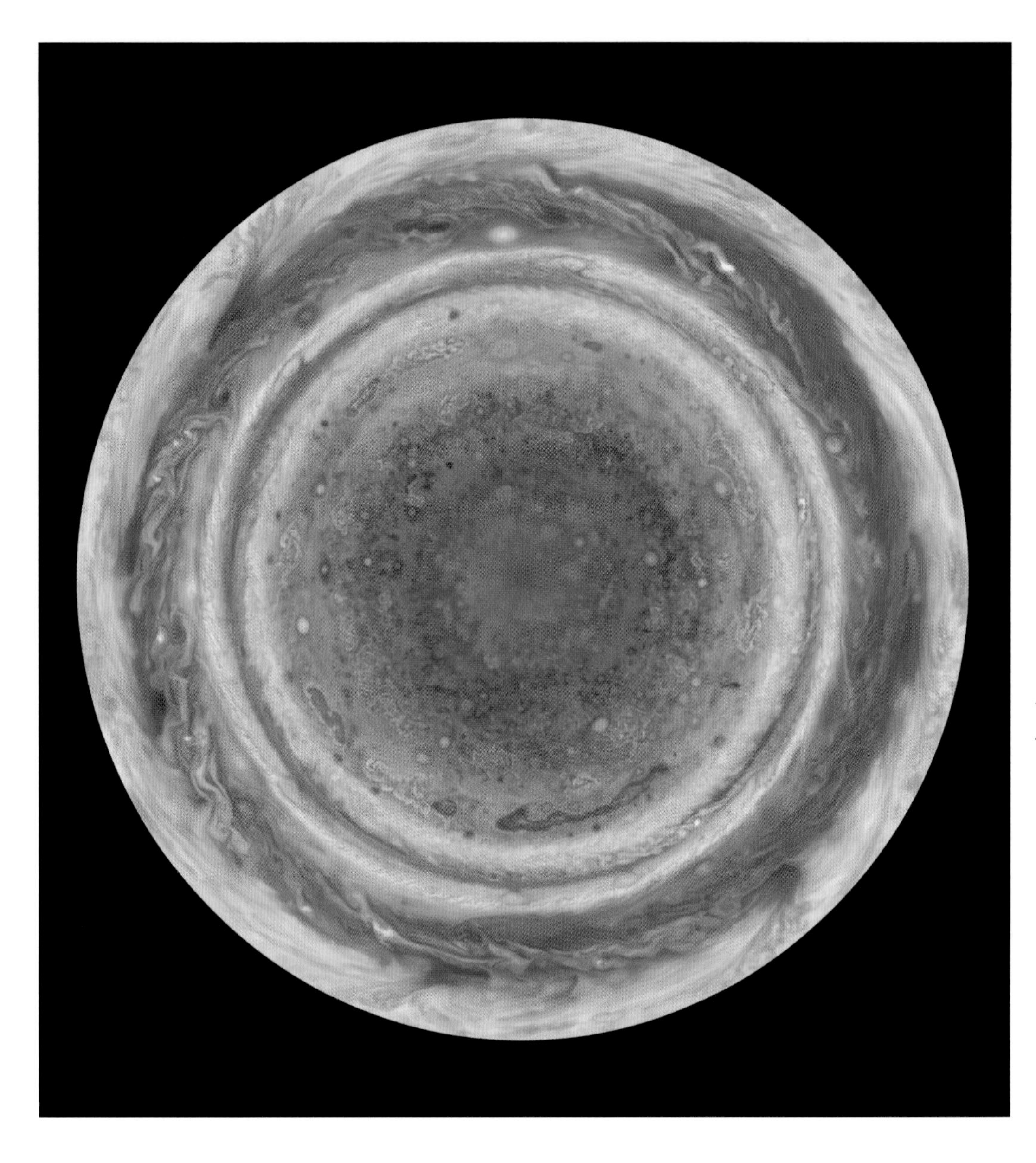

Jupiter viewed from above by the Cassini spacecraft in 2000. The north pole is central, and the equator is around the circumference of the image. Bands of cloud and stormy vortices are clearly visible.

Galileo could not have known that these planets were made of gas. Indeed, the concept of gas didn't really exist at the time; the discovery that there are different types of 'air' still lay in the future. Scientists calculated that Jupiter and Saturn are made largely of gas by working out their mass from the orbits of their moons, and their density from their size, using the calculated mass. The term 'gas giant' was first used by James Blish in a science fiction story written in 1952.

MAPPING GAS GIANTS

Many of the techniques used to map rocky planets don't work with gas giants. We can't bounce rays off their surface to determine their terrain as they have neither surface nor terrain. Their albedo (brightness) doesn't relate to mineral composition, and we can't send rovers crawling over them to take samples. We can use spectrographic techniques to work out what they are made of and to measure their temperature, and we can look at them in visible light, infrared and ultraviolet, but seeing through the cloud layers in any way is a challenge.

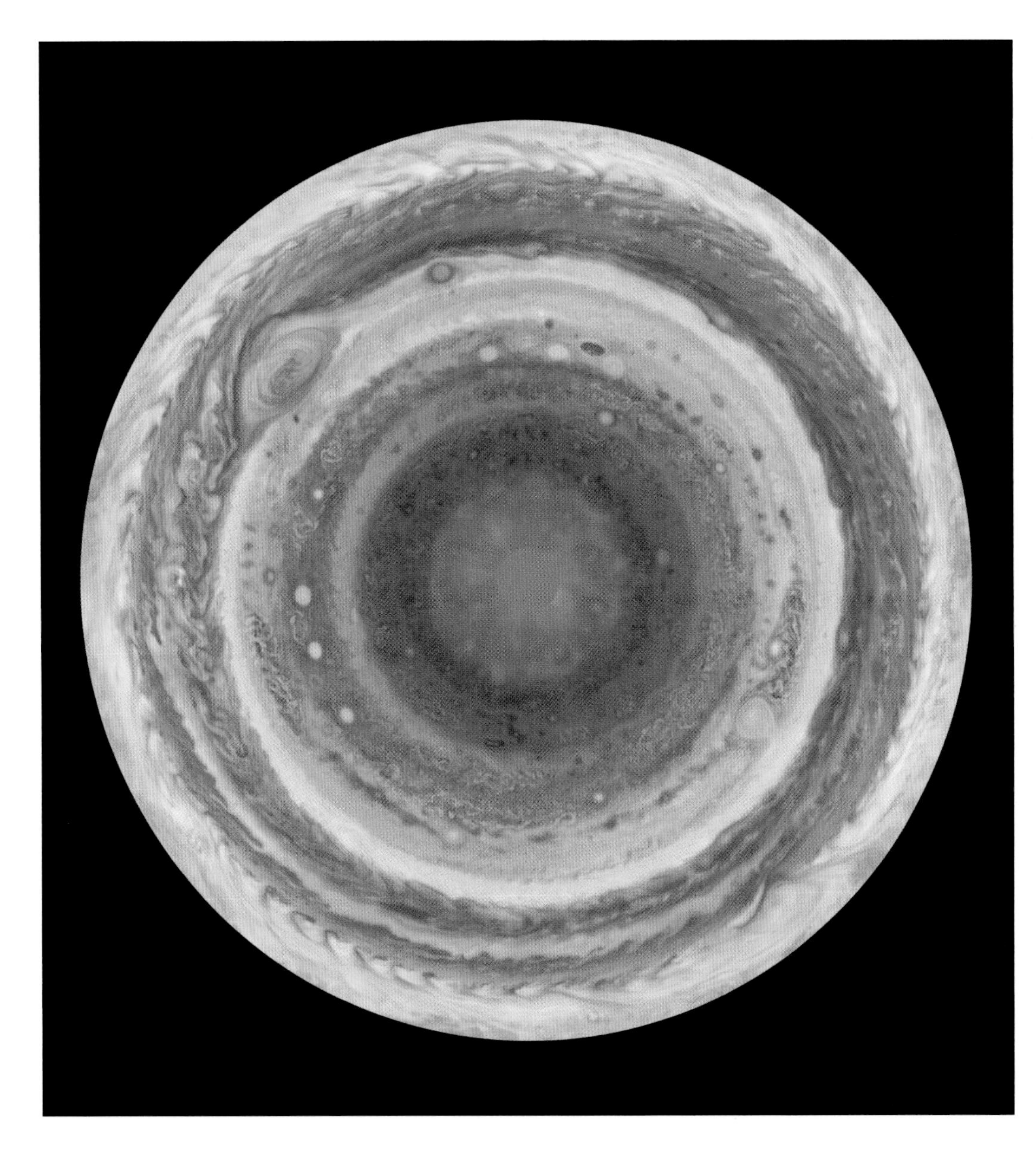

Jupiter viewed from beneath by the Cassini spacecraft in 2000. The south pole is central and the equator is around the circumference of the image. The red spot on the upper left is a large storm; smaller storms are visible swirling in other bands of cloud.

There are no fixed features to help us distinguish one side of a gas planet from another. The clouds and storms are themselves moving around the planet in the same way as these features do on Earth. Mapping a gas planet would be like trying to map Earth without making any reference to land and sea. The only fixed points of the gas giants are the poles.

CLUSTERING MOONS

The gas giants and the ice giants beyond them have many moons. Some of these probably formed at the same time as the planets and are considered 'regular' moons, but others have been captured – dragged into the planet's orbit. These 'irregular' moons often have retrograde orbits, so they circle the planet in the opposite direction to its own rotation. Many are so small and dark that they have only been detected by visiting spacecraft. Yet the larger moons of the gas giants are among the most likely places in the solar system to be habitable. Though distant and cold, some have underground oceans that could provide the kind of environment in which Earth's early microbial life flourished.

JUPITER

After Earth, Jupiter is the planet with the most interesting and dynamic appearance. Its lively atmosphere provides an ever-changing pattern of striped bands, swirls and whorls as the clouds and storms rip around the upper atmosphere.

A colour-enhanced image of Jupiter from photographs taken by NASA's Juno spacecraft in February 2019. A zone of white cloud circles the planet. The Great Red Spot is visible on the right.

Jupiter's size relative to that of Earth.

Planetary focus	Jupiter
Length of year	4,333 days (12 years)
Length of day	10 hours
Size x Earth mass	318
Size x Earth radius	11
Average distance from Sun	778 million km (483 million miles)
Moons	79
Discovered	Prehistory
Missions	Flyby: Pioneer 1 and 2 (1973, 1974), Voyager 1 and 2 (1979) Gravity assist: Ulysses (1992, 2000), Cassini-Huygens (2000), New Horizons (2007) Orbiter: Galileo (1995–2003), Juno (2016–)

Jupiter is by far the largest planet in the solar system, with twice the mass of all the other planets combined. After the formation of the Sun, Jupiter took most of the remaining mass and evolved in the same way and with about the same composition as a star. It cannot become a star, however, as its mass is too low to produce sufficient pressure at the core to start nuclear fusion.

Within Jupiter, hydrogen becomes liquid, giving the planet an ocean of liquid hydrogen 20,000 km (12,400 miles) deep, the largest in the solar system (although an ocean of hydrogen is unlike any ocean on Earth). Eventually, deep inside Jupiter, the atmospheric pressure becomes so great that the hydrogen fluid starts to behave like a metal, conducting electricity and turning the planet into an enormous generator. This, combined with Jupiter's rapid rotation, creates a magnetic field 725 million km (450 million miles) long.

Jupiter's core might be solid or a hot super-dense fluid. It is probably composed of iron and silicate rocks such as quartz and is likely to be as large as Earth.

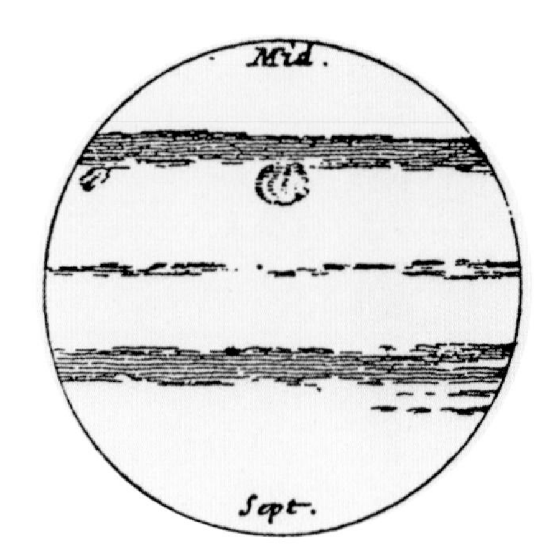

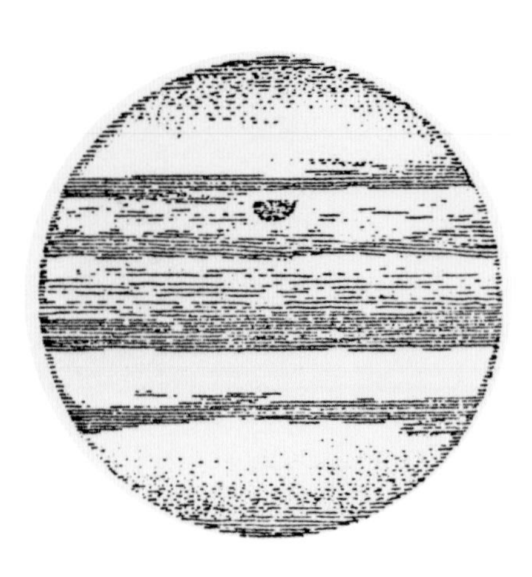

Cassini's sketches of Jupiter made in 1672 (left) and 1677 (right) show changes in the planet's appearance. South is at the top in both images.

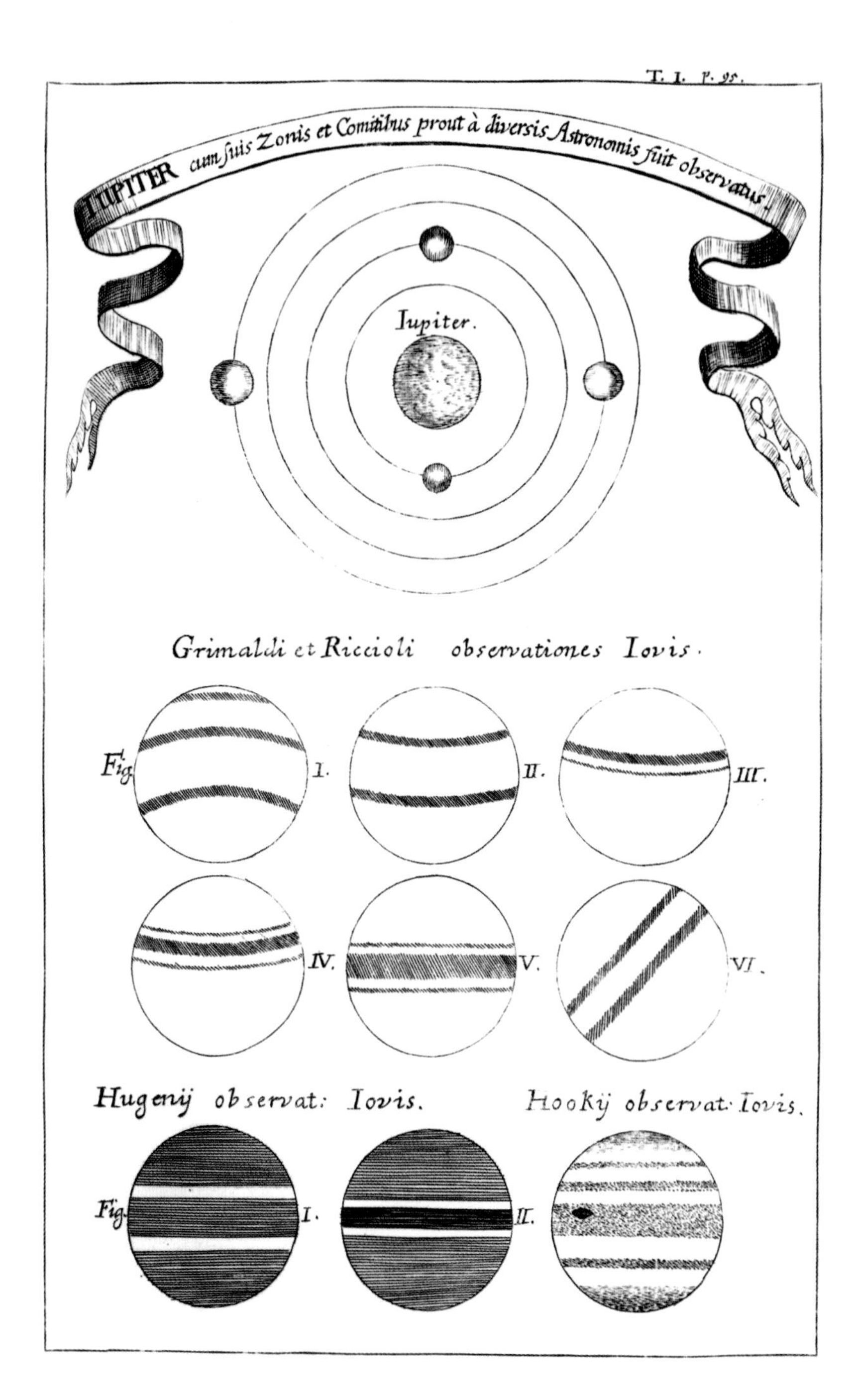

EARLY VIEWS OF JUPITER

When Galileo saw Jupiter, his telescope wasn't powerful enough to make out any spots on its surface. The spots we see now are vast, high-speed storms, whirling vortices of winds in the upper atmosphere that can last decades or centuries.

A large spot was first described on Jupiter by English scientist Robert Hooke in 1664, although it is unlikely it was the same Great Red Spot that we see now. Giovanni Cassini reported seeing a spot through his telescope a year later. It's not clear whether Cassini's spot, in the southern hemisphere, was the same one that Hooke saw because their accounts don't match. Cassini's was documented and drawn consistently from 1665 until 1713, after which no spot is mentioned until 1830. The gap in observations of 117 years suggests the original spot seen by Cassini disappeared and another eventually formed.

Left: *Observations of Jupiter published by Zahn in 1696. At the top of the image, Jupiter is shown surrounded by the four Galilean moons; at the bottom are depictions of the planet's bands of cloud by astronomers Francesco Grimaldi, Giovanni Riccioli, Christiaan Huygens and Robert Hooke.*

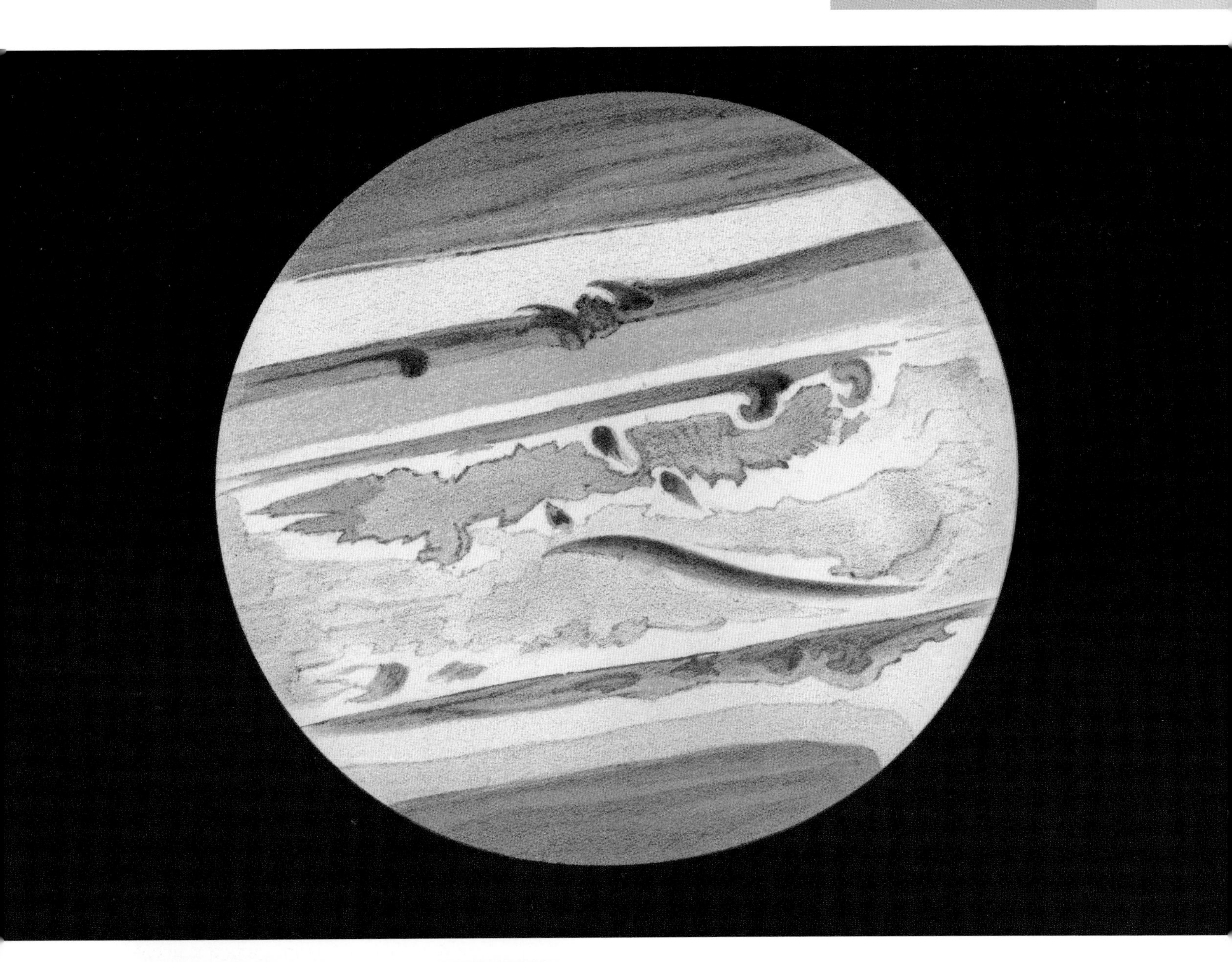

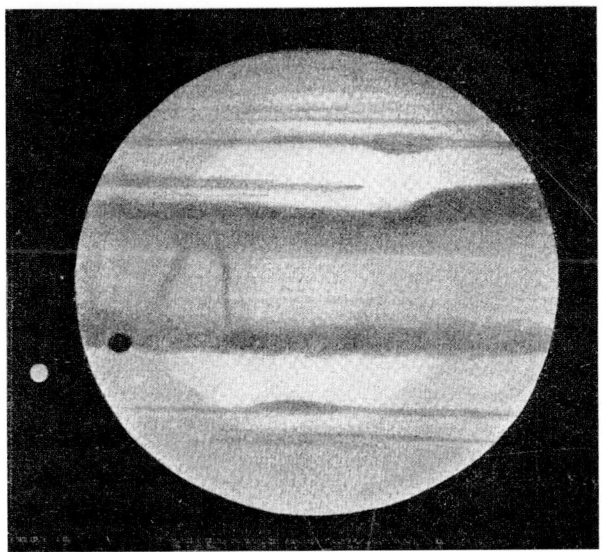

Above: *Jupiter drawn and coloured from the observations of Pietro Tacchini in 1873.*

Left: *Jupiter with one (unnamed) moon casting a shadow on the planet's atmosphere, 1910.*

A VERY STORMY STORM

By the time the Great Red Spot was next mentioned, in the 19th century, telescopes had improved considerably. By 1880, its size was calculated at about 48,000 km (30,000 miles) across or nearly four times the width of the Earth. It's currently about 1.3 times the diameter of Earth, but it is still shrinking and its winds are probably slowing. Its changing size is shown in this photograph from 2015 (right) alongside a reconstructed approximation of how it looked in 1890 (left).

The Great Red Spot is now known to be a storm caught between two gulf streams, a very strong westward stream to the north and a much weaker eastward one to the south. The winds within it travel counter-clockwise, as calculated in 1966 and confirmed by the first time-lapse video returned by the Voyager craft. Wind speeds at the edge are around 432 km (269 miles) per hour, but in the centre there is little movement. The cloud tops of the storm are about 8 km (5 miles) high.

VOYAGER 1, JUPITER'S GREAT RED SPOT, 1979

This false-colour image was made from three black-and-white negatives taken by Voyager 1. Vibrant bands of fast-moving cloud continuously circle Jupiter's atmosphere. This is the first detailed view of the Great Red Spot and the patterns of its surrounding belt of cloud. No comparable view can be achieved from Earth.

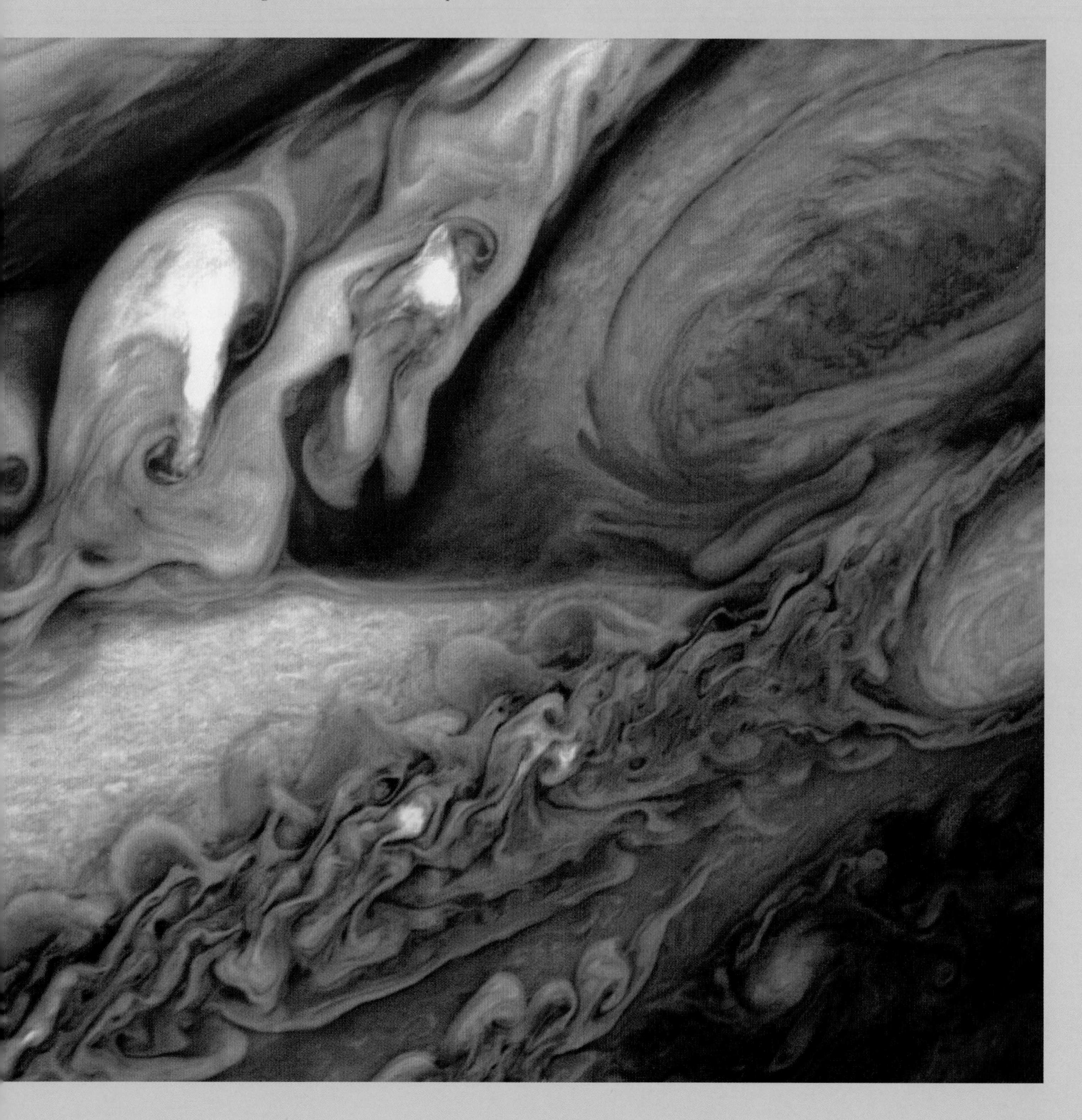

JUNO, CYCLONIC STORM, 2019

This false-colour image of a cyclonic storm over Jupiter's northern hemisphere was created from a photograph taken by the Juno spacecraft in 2019 from 8,000 km (5,000 miles) above the cloud tops. The white clouds are made of water and ammonia ice and can form towers 50 km (30 miles) high.

JUNO, CLOUDS OVER JUPITER'S SOUTH POLE, 2017

The Juno mission arrived near Jupiter in 2016 and has been orbiting the planet, taking photographs and measurements of the atmosphere. The image below of Jupiter's south pole reveals localized swirling squally storms which contrast with the large bands circling the wider parts of the planet. From Juno's measurements of gravity variations, it's now thought that the turbulent storm bands may extend down at least 2,993 km (1,860 miles). At that depth, the stormy atmosphere would account for around 1 per cent of the planet's mass, or three times the mass of Earth. In contrast, Earth's atmosphere accounts for less than one millionth (0.0001 per cent) of our planet's mass.

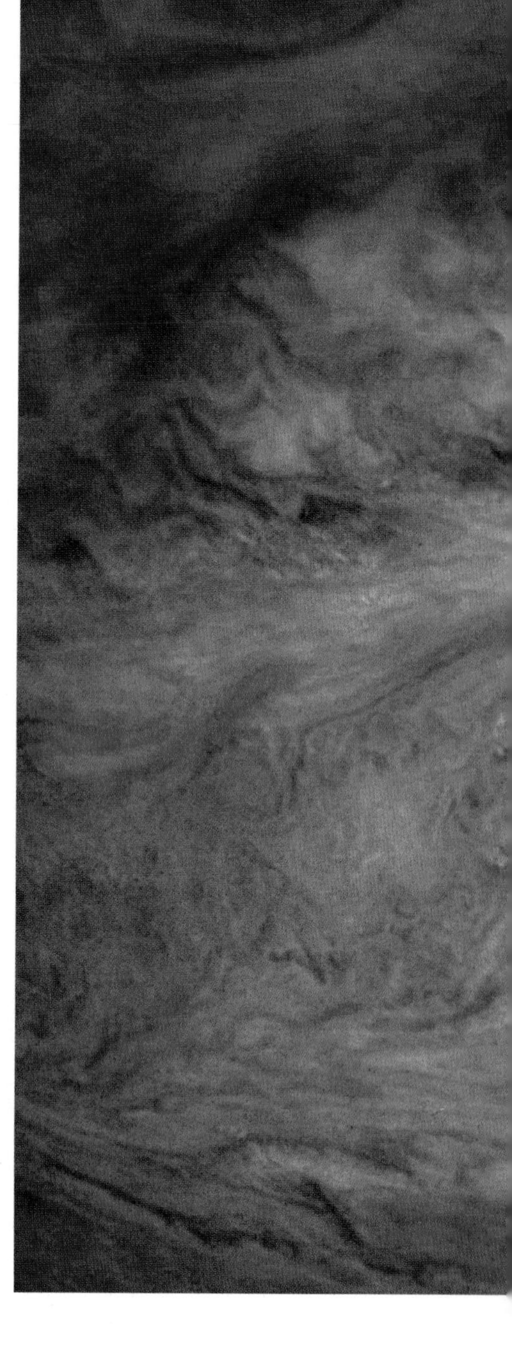

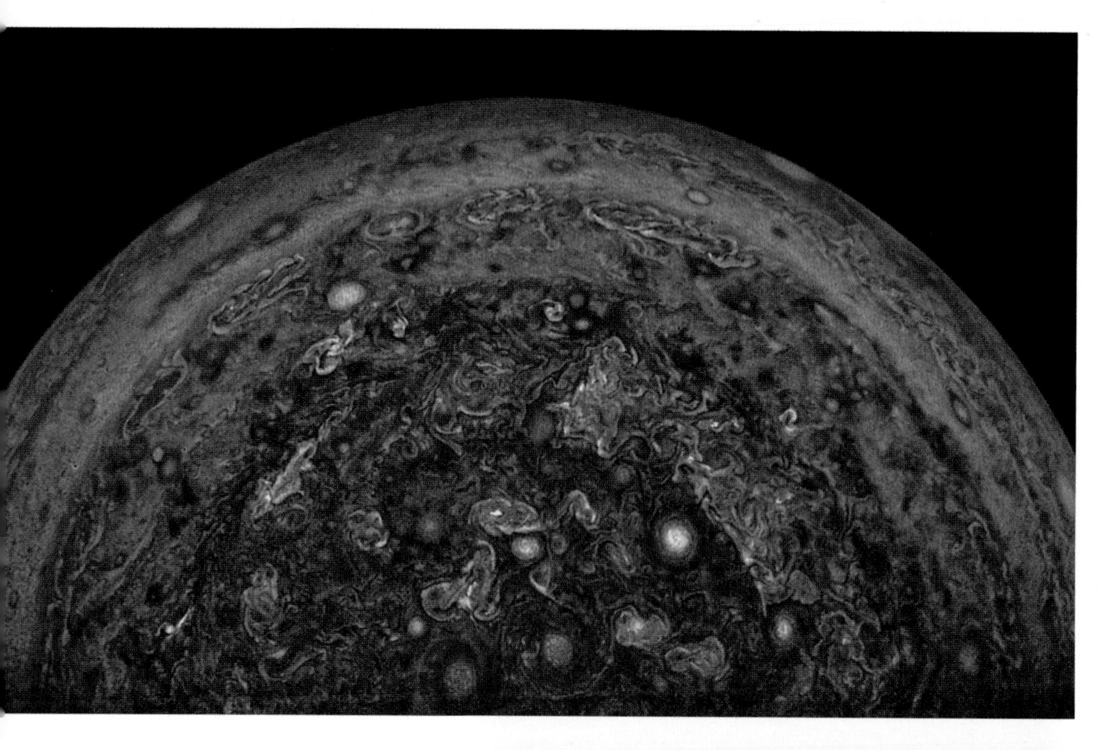

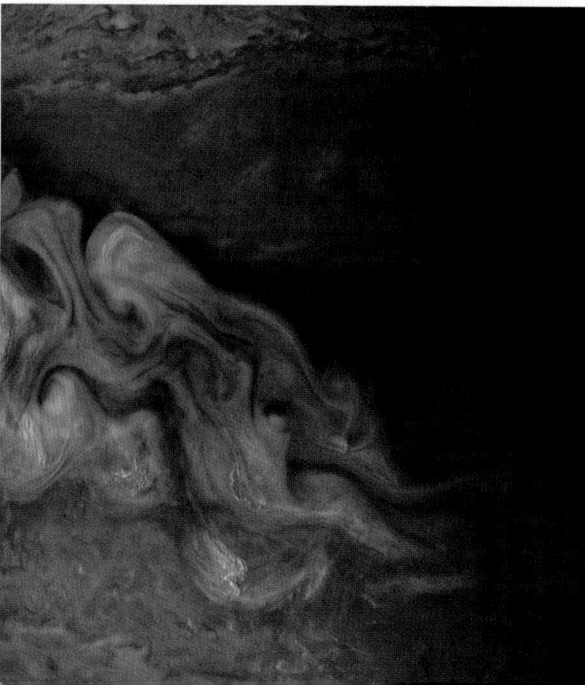

Right: *A close-up, enhanced-colour image of Jupiter's clouds.*

JUNO, DETAIL OF CLOUDS, 2017

Detailed examination of the cloud patterns of Jupiter found by Juno revealed that models of the planet's weather systems need to be revised. The polar areas are covered with cyclones and anti-cyclones, some of them larger than Earth. In this enhanced-colour image, white clouds are the highest, composed of water/ammonia ice; brownish clouds contain ammonium hydrosulphide and lie in a lower layer. Blue areas indicate deep cloud, below the brown and white layers.

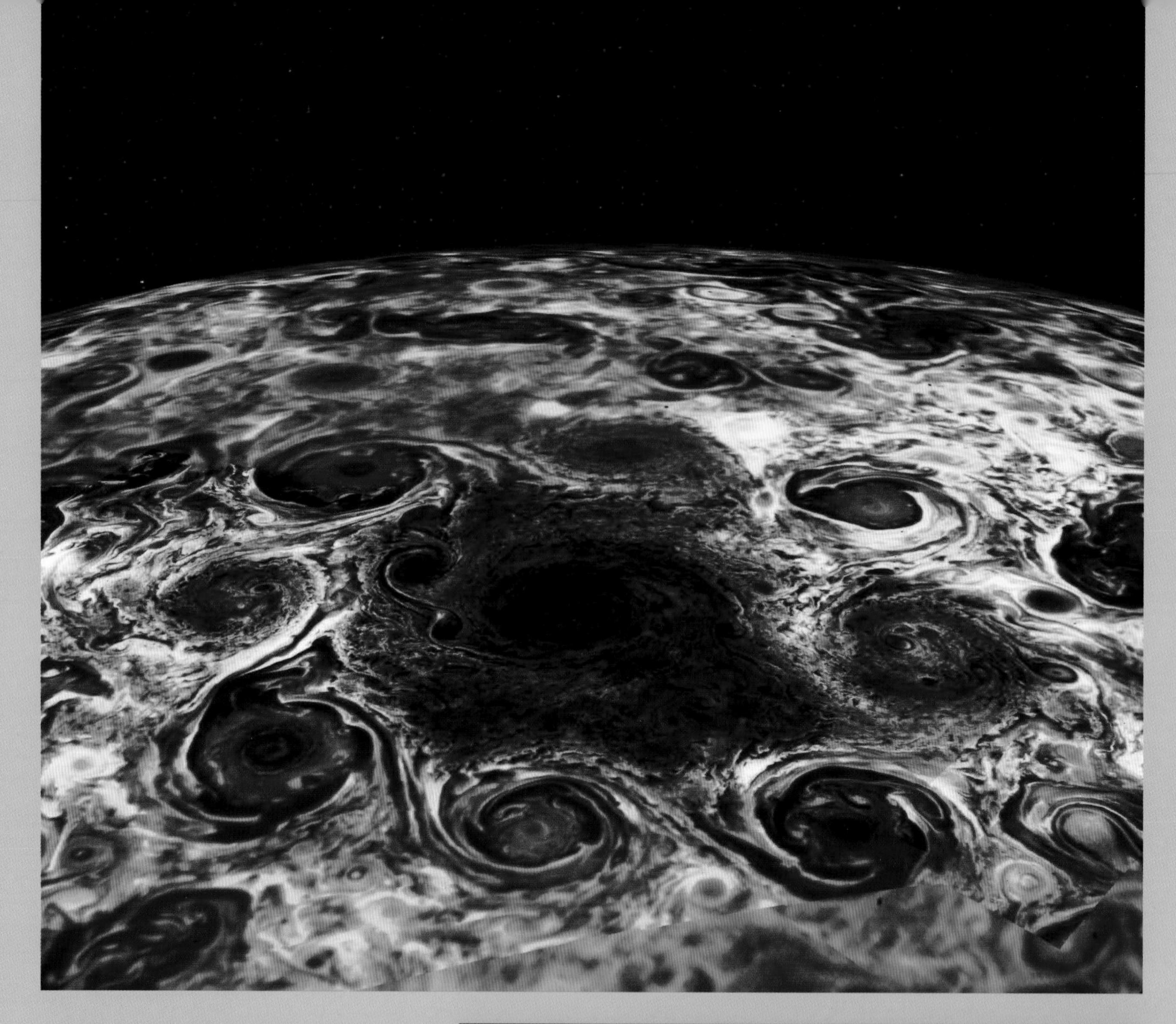

JIRAM, NORTH POLE IN INFRARED, 2018

This image of Jupiter's north pole was built up from images taken by the Juno mission's Jovian Infrared Auroral Mapper (JIRAM). The central cyclone over the north pole is circled by eight smaller cyclones, with winds up to 354 km (220 miles) per hour. At the south pole, there are five smaller cyclones around the central one. Darker areas are of thicker cloud and colder, -118° C (-181° F); lighter areas are of thinner cloud and warmer, -13° C (9° F). Heat is probably generated fairly evenly within the planet, so the differences in temperature at the top of the cloud probably reflect differences in the density of clouds blocking that heat.

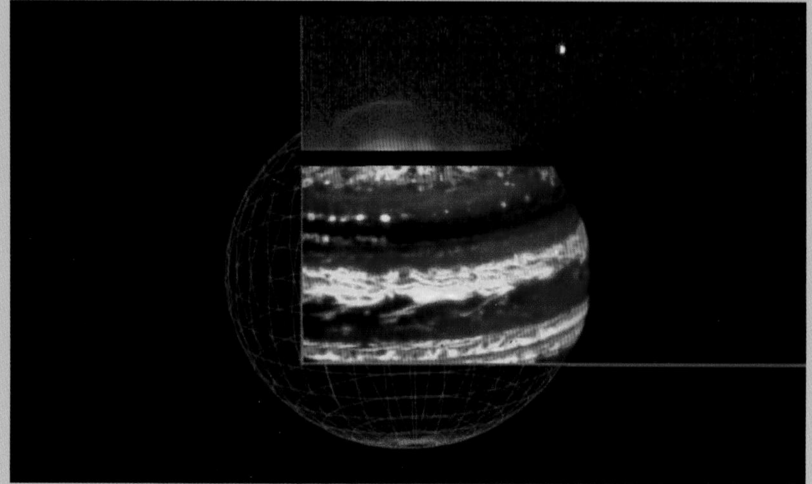

JIRAM, JUPITER IN INFRARED, 2016

Detecting emissions from Jupiter at different wavelengths, JIRAM can reveal both the temperature of the clouds (shown in the red/yellow part of the image) and the aurora coming from the planet (in the blue/purple part of the image). Jupiter's moon Io is visible at the top right.

JUNO, WITHIN THE CLOUDS, 2017

This image shows a cross-section of the outer part of Jupiter's atmosphere, which Juno's microwave radiometer measured from the cloud tops to 350 km (220 miles) below. The planet's atmosphere probably has three bands of cloud made up of tiny ice crystals of different substances at different altitudes. The highest clouds are ammonia ice, the middle clouds are ammonium hydrosulphide crystals, and the lowest are water ice. The vivid colours are probably produced by plumes of warmer gases containing phosphorous and sulphur rising from further down in the planet.

Scientists expected differences to level out at around 100 km (60 miles), but there is no uniform layer even at the deepest level that microwaves can reveal. In the illustration on the right, orange signifies high levels of ammonia and blue shows low levels of ammonia. There is a band of high ammonia concentration near the equator and an area of low ammonia just north of it.

NASA, MAGNETIC FIELD, 2018

Jupiter has long been known to have the strongest magnetic field of any planet in the solar system. It extends up to 7 million km (4 million miles) towards the Sun and as far as Saturn in the opposite direction. It is produced by movement in the planet's metallic hydrogen outer core. Juno has shown Jupiter's magnetosphere to be more intense and less regular than expected. Its lumpiness suggests it might be generated closer to the nominal surface of the planet than previously thought. The liquid ocean of metallic hydrogen above Jupiter's core is believed to act like a dynamo, as electricity flows through it and the planet turns on its axis.

Above: *An artist's impression of Juno in orbit around Jupiter.*

Below: *Jupiter's magnetic field.*

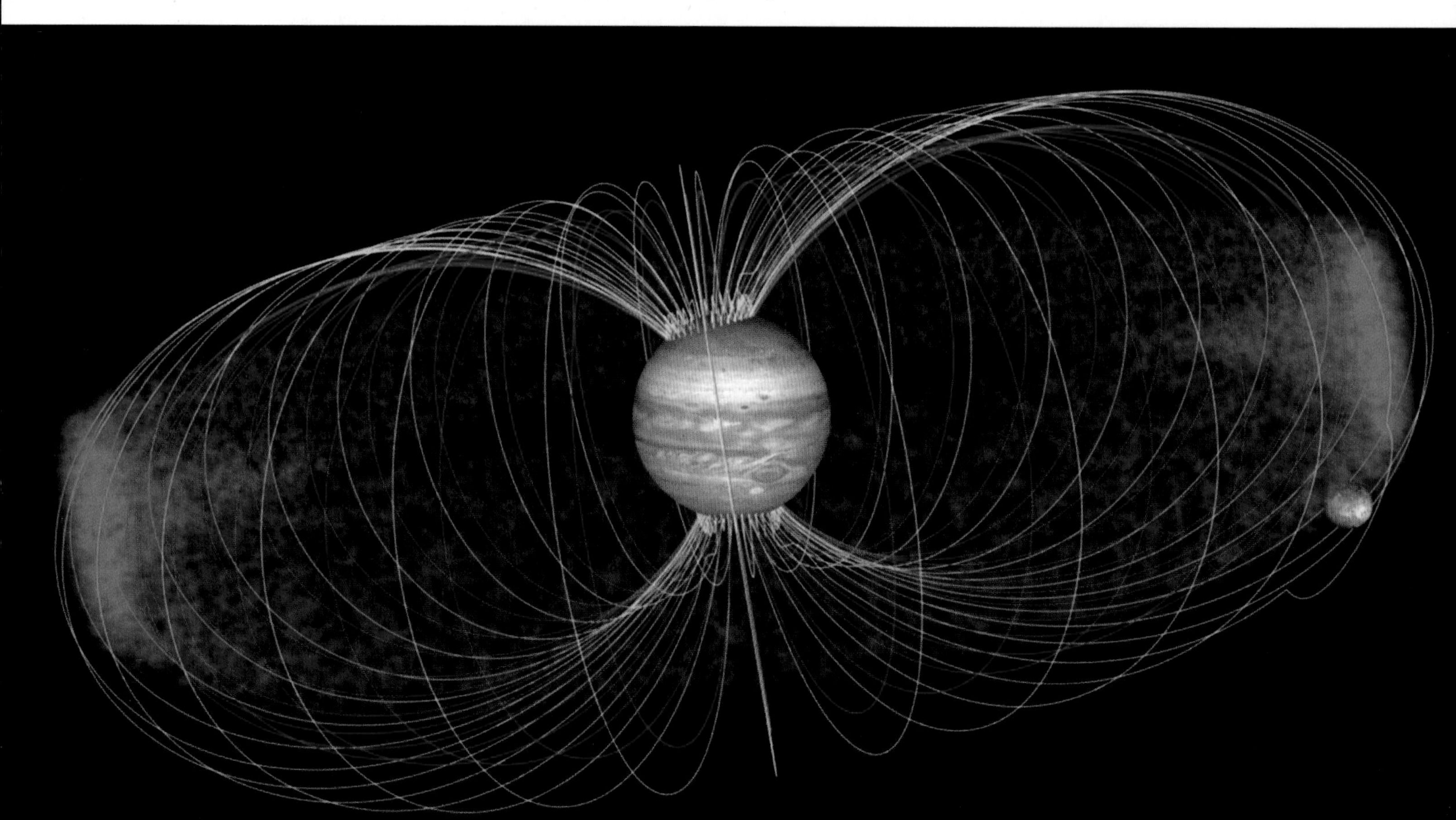

NASA & ESA, JUPITER'S NORTHERN AND SOUTHERN AURORA, 2007/2016

These images of the northern and southern lights (aurorae) of Jupiter have been created from data gathered by NASA's Chandra X-ray and ESA's XMM Newton observatories. The aurora of the north pole is shown in the top image, and that of the south pole in the lower image. The X-ray hot-spots produced, visible in pink, can cover an area half the size of Earth's surface.

The aurorae at the two poles are independent of each other (unlike Earth's aurorae, which mirror each other). At the south pole, X-ray emission follows a regular pattern of pulses at intervals of 11 minutes, but emission at the north pole is erratic. Astronomers are investigating the theory that the aurorae are caused by interaction between Jupiter's magnetic field and the solar wind producing magnetic waves. Particles would then 'surf' the magnetic waves, gaining energy. Collisions between high-speed particles and the atmopshere would produce bursts of X-rays. The mechanism for accelerating the particles sufficiently is not known, nor is the impact of very high speed particles colliding with the planet's poles.

JUPITER'S MOONS

Jupiter has 79 moons, more than any other planet in the solar system. When Galileo spotted the four largest moons, he became the first person ever to see moons orbiting a planet other than Earth. These account for most of the moon-mass of the system; all Jupiter's other moons combined account for just 0.003 per cent of the moon-mass. At 5,268 km (3,273 miles) across, Ganymede is the largest moon in the solar system – larger than the planet Mercury. At the other extreme, Jupiter has seven tiny moons with a diameter of just 1 km (0.6 miles) each.

This image taken by the Hubble Space Telescope in 2015 shows the transit of three of Jupiter's larger moons, Io (top right), Callisto and Europa (lower left). The first two moons cast black shadows onto the cloud tops; the shadow of Europa is not visible here.

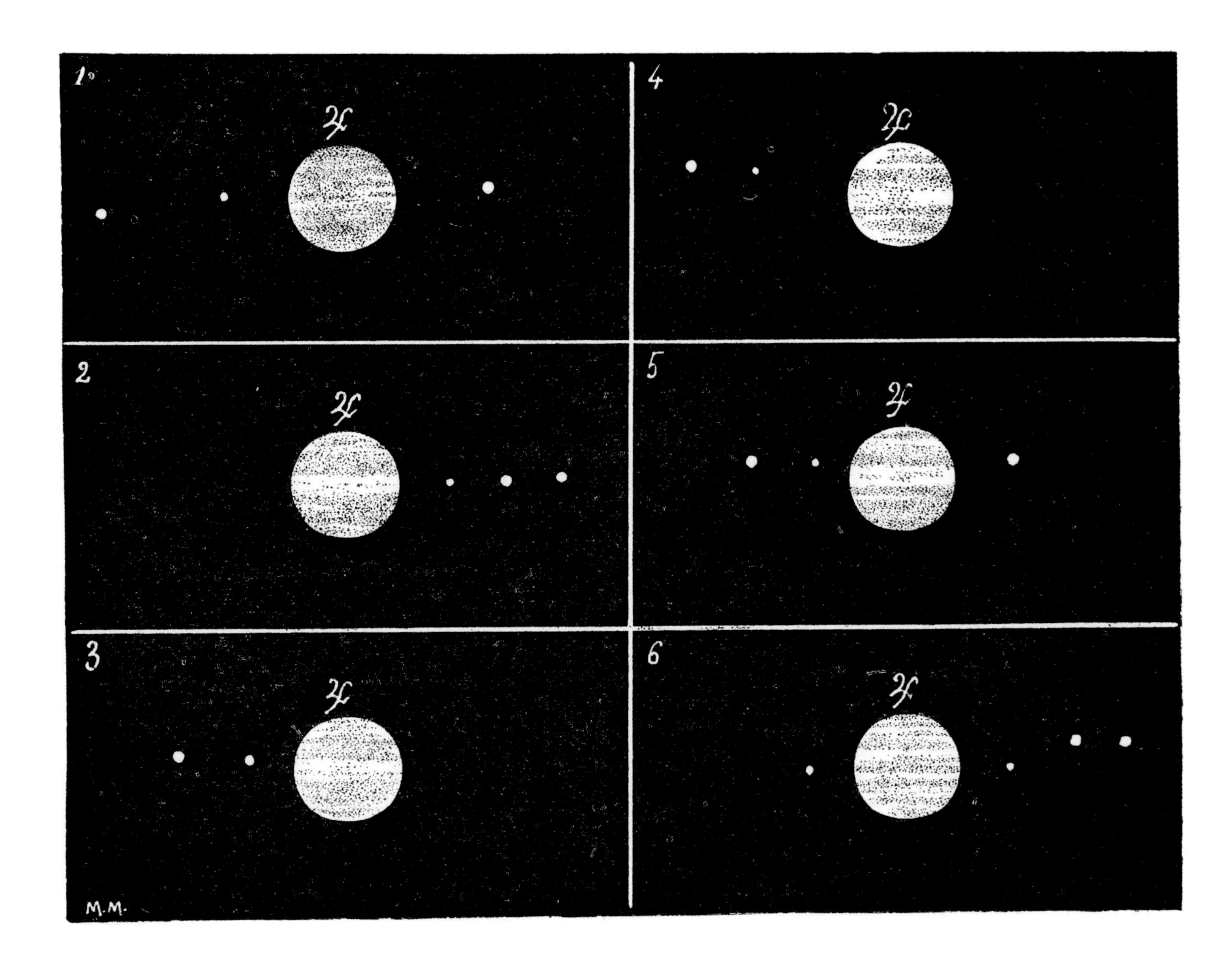

GALILEO, MOONS OF JUPITER, 1610

Galileo plotted the locations of the bright spots he saw around Jupiter repeatedly until he concluded that the planet has four satellites. At first he wanted to name them in honour of his patron, Cosimo de' Medici. Their current names – Ganymede, Io, Europa and Callisto – were given by the German astronomer Simon Marius, who discovered the moons around the same time as Galileo. Rejecting these, Galileo instead named the moons using Roman numerals, starting from Jupiter and numbering outwards. His scheme was used until the mid-20th century, when the discovery of moons between the numbered ones made the system untenable. At this point, Marius' names were adopted.

NASA, GALILEAN MOONS, 1997

This composite image of the Galilean moons and part of Jupiter shows the relative sizes of (from top) Io, Europa, Ganymede and Callisto. The image of Callisto was taken by Voyager in 1979; the others were taken by the Galileo craft in 1996. The smallest moon, Europa, is about the same size as Earth's Moon.

No moons were found between Galileo's discovery and 1892, when the American astronomer Edward Barnard discovered Almathea. By the time Voyager 1 flew past Jupiter in 1979, thirteen moons had been discovered (plus one, Themisto, which had been seen but wasn't confirmed until 2000). Voyager found three more moons close to Jupiter. No more appeared for twenty years, then 34 were found between 1999 and 2003. More appeared slowly, and then a clutch of ten in 2018, bringing the total to 79.

Only eight of Jupiter's moons are 'regular satellites' that formed in orbit around the planet. The remainder have been captured and dragged into its orbit. As of 2019, 27 moons are still unnamed. The outer moons all have eccentric and inclined orbits (the plane of their orbit is at an angle to the plane defined by Jupiter's equator).

The inner moons are the four Galilean moons; four small moons which lie even closer to the planet are known as Metis, Adrastea, Amalthea and Thebe. Adrastea was the first moon to be discovered from photographs taken by a spacecraft (Voyager 2 in 1979). All four are very small and irregular in shape. Amalthea is the reddest object in the solar system and is a surprisingly low-density moon, composed of ice and rock rubble.

NASA, IO, 1999

The closest of Jupiter's large moons, Io, is the most volcanically active body in the solar system. Its 400 known volcanoes throw out material to a height of more than several dozen kilometres, providing the sulphur dioxide that makes up Io's thin atmosphere. Deposits of yellow, white, grey and brown sulphurous volcanic products cover other parts of the moon, easily visible in this true-colour image. The reddish and darker areas are most recently volcanically active.

The moon is tidally locked to Jupiter, so the same side always faces the planet, but Europa and Ganymede also exert a considerable gravitational force on it, producing a highly elliptical orbit. The result is strong tidal forces pulling on the rock of Io and causing it to bulge with 'tides' of up to 100 m (330 ft) in its solid surface.

GALILEO, VOLCANOES ON IO, 2001

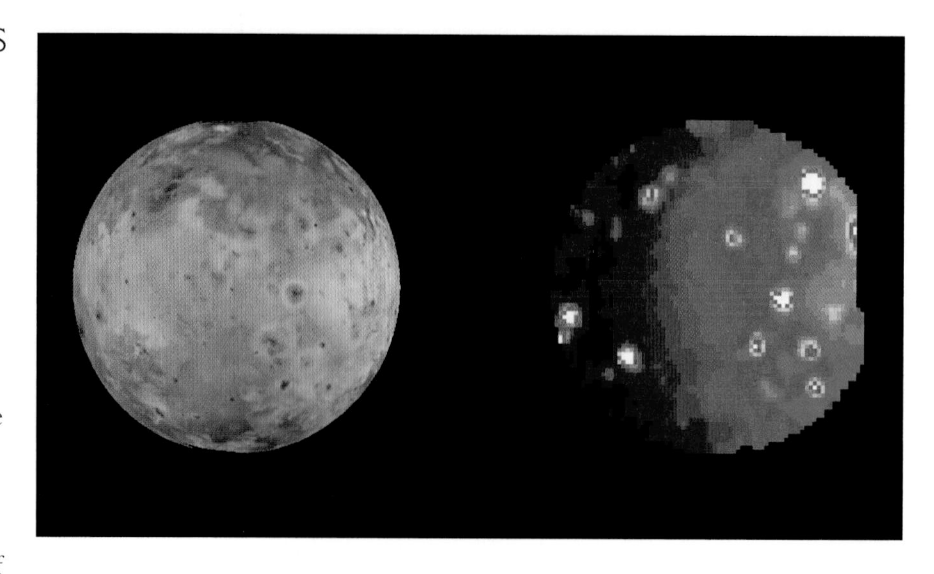

An infrared map of Io (right) makes it easy to see the location of volcanoes and match them to geological features on the surface (left). The hottest regions are in red, yellow and white; the coldest are in blue.

Io is the most volcanically active world in the solar system. Four previously unknown volcanoes were discovered using this image.

The tidal forces generate a lot of heat within Io, keeping the rock just below its surface molten and prone to burst out of volcanoes at any opportunity. There are lakes of liquid lava on the surface. The crust of Io is constantly renewed by this volcanic activity, so there are few impact craters that have not been flooded with lava.

This composite and colour-enhanced image is made from a mosaic of photos taken by the Galileo craft in 1999. It shows an ongoing volcanic eruption at Tvasthtar Catena, *a chain of volcanoes in the northern hemisphere of Io.*

GALILEO, EUROPA, 2000

Galileo photographed Europa (left) on a flyby in 1996, revealing a surface of solid ice with extensive cracks. The surface temperature is -160 °C (-260 °F) at the equator and -220 °C (-370 °F) at the poles.

Photos from the Hubble Space Telescope released in 2016 show plumes of water erupting from the surface and spurting 200 km (125 miles) into space. This supports the theory that there is an ocean of water 10–30 km (6–19 miles) beneath the ice crust. The ocean could be 100 km (60 miles) deep, in which case its total volume would be two to three times that of all of Earth's oceans. This hypothetical ocean is considered one of the more likely places to find life elsewhere in the solar system.

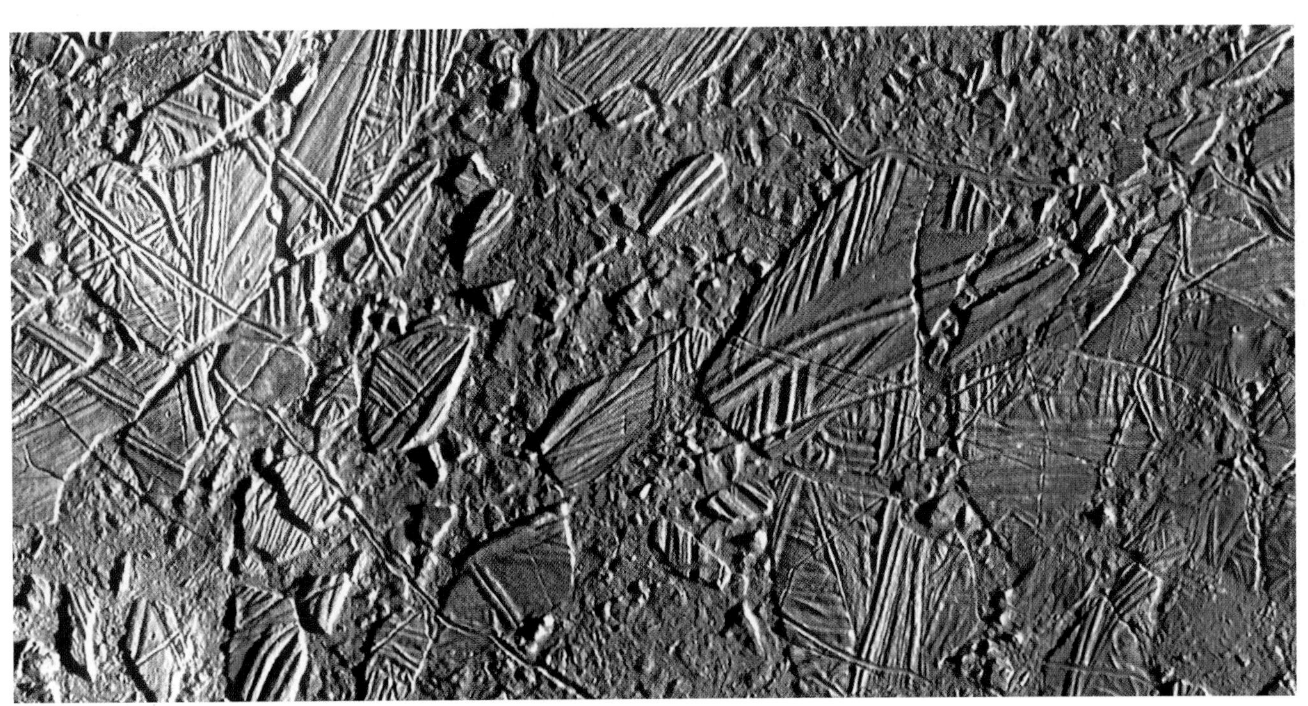

GALILEO, ICE ON EUROPA, 1996–7

This disrupted ice crust in the Conamara region of Europa shows both bright white and blue areas that were dusted with ice particles at the formation of the Pwyll crater 1,000 km (621 miles) to the south. The reddish-brown colouring here and elsewhere on Europa is thought to be the result of minerals and sulphur compounds from within the planet, brought to the surface through cracks and impact craters.

GALILEO AND VOYAGER, GEOLOGY OF GANYMEDE, 2014

Ganymede was first viewed by Voyagers 1 and 2, and later observed by the Galileo mission. This mosaic of its surface was created from images taken by the Voyager (1979) and Galileo (1996) missions. The surface of Ganymede is ice and silicate rock. The low-albedo, darker terrain is around four billion years old; rich in clays and heavily cratered, it covers 40 per cent of the moon. The lighter terrain is more recent (but still very ancient); it has far fewer craters and is criss-crossed by grooves that were probably caused by tension within the moon, or water released from beneath the surface. The grooves have ridges up to 700 m (2,000 ft) high and run for thousands of kilometres over the surface.

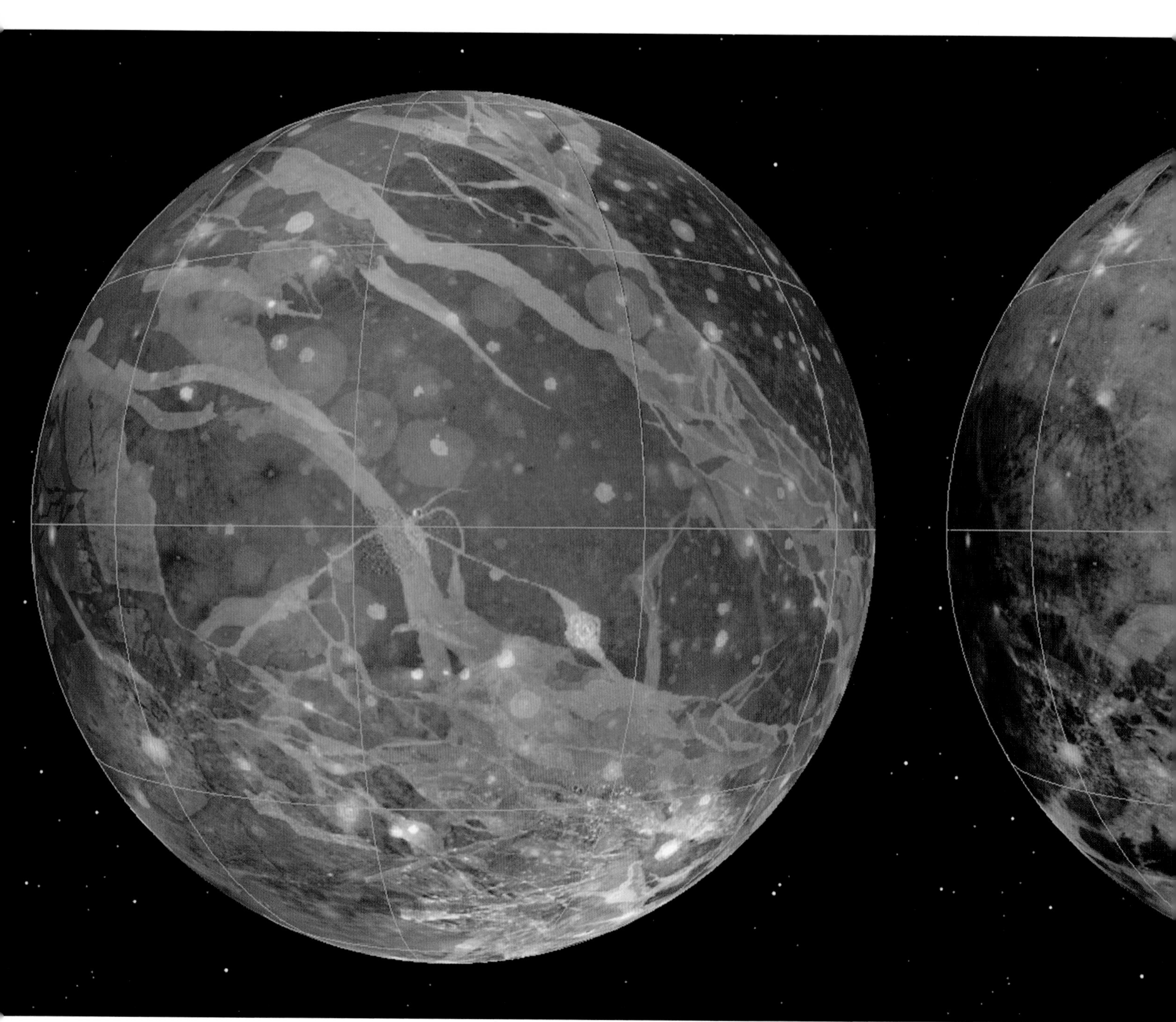

Ganymede has a liquid iron core that takes up to half of its diameter. This is surrounded by a mantle of silicate rock and then a layer of ice and possibly water 800 km (500 miles) thick. If there is a liquid ocean, it's thought to be sandwiched between two layers of ice, one directly above the rocky mantle, the other at the surface. This ocean might be the largest in the solar system.

The geological map on the left shows the distribution of different types of terrain. The oldest, cratered areas are reddish brown; the younger areas are blue if grooved and blue-green if smooth. Areas shaded purple are a mixture of grooved and smooth terrain.

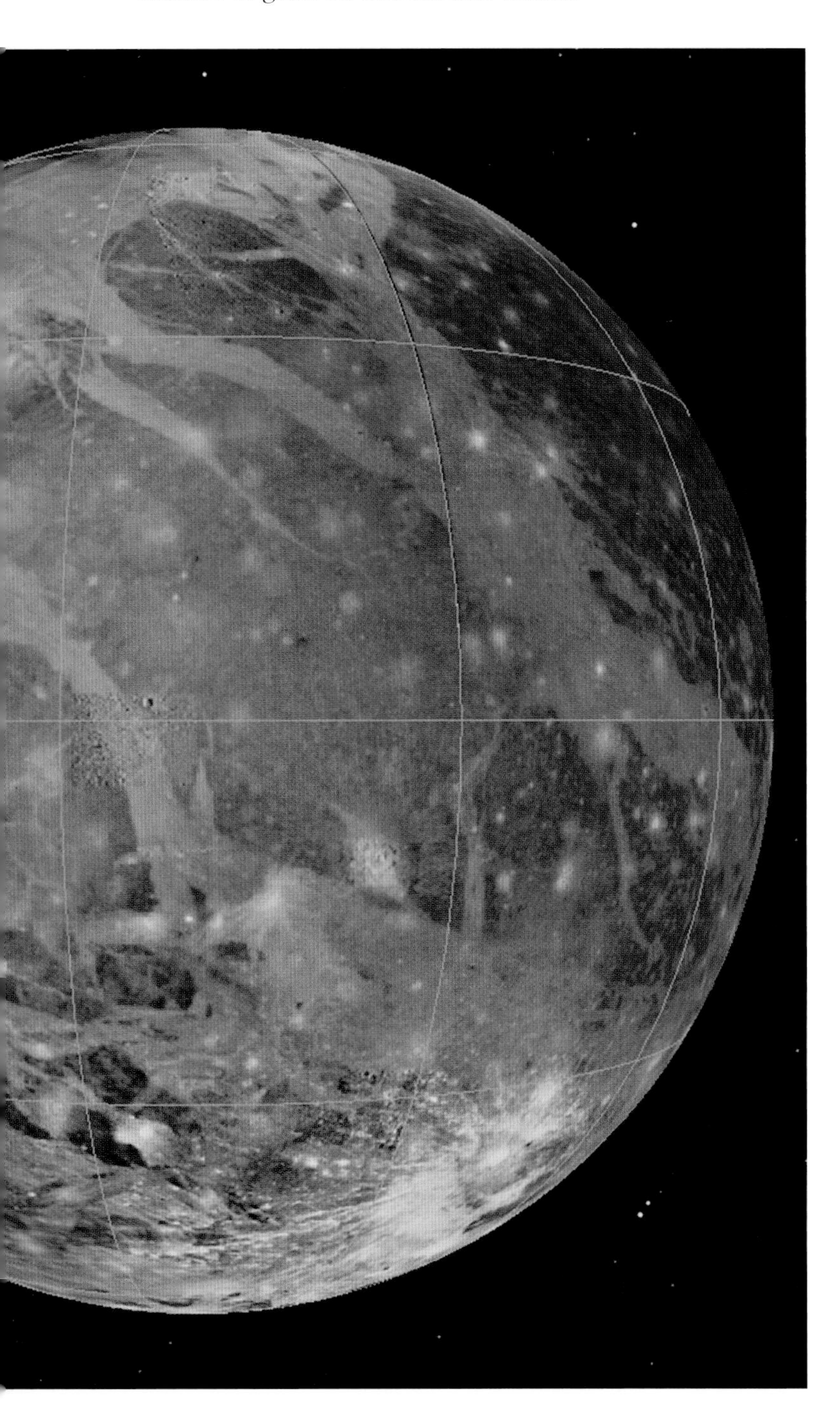

GALILEO, CALLISTO, 2001

Callisto (right) is Jupiter's second-largest moon, about the same size as Mercury. It's believed to harbour an ocean of salt water as deep as 250 km (155 miles) below the surface, making it a potential host for some form of life. Beside the ocean, the planet is likely to be a mix of ice, rock and metal, perhaps all the way to the centre, with no differentiated core.

Callisto is the most heavily cratered body in the solar system. There is no tectonic activity or weathering to wear away the craters, though some seem to be capped with water ice which makes the shining spots visible on the surface.

While the innermost moons orbit Jupiter in a matter of hours, Callisto takes seventeen days to travel around the planet.

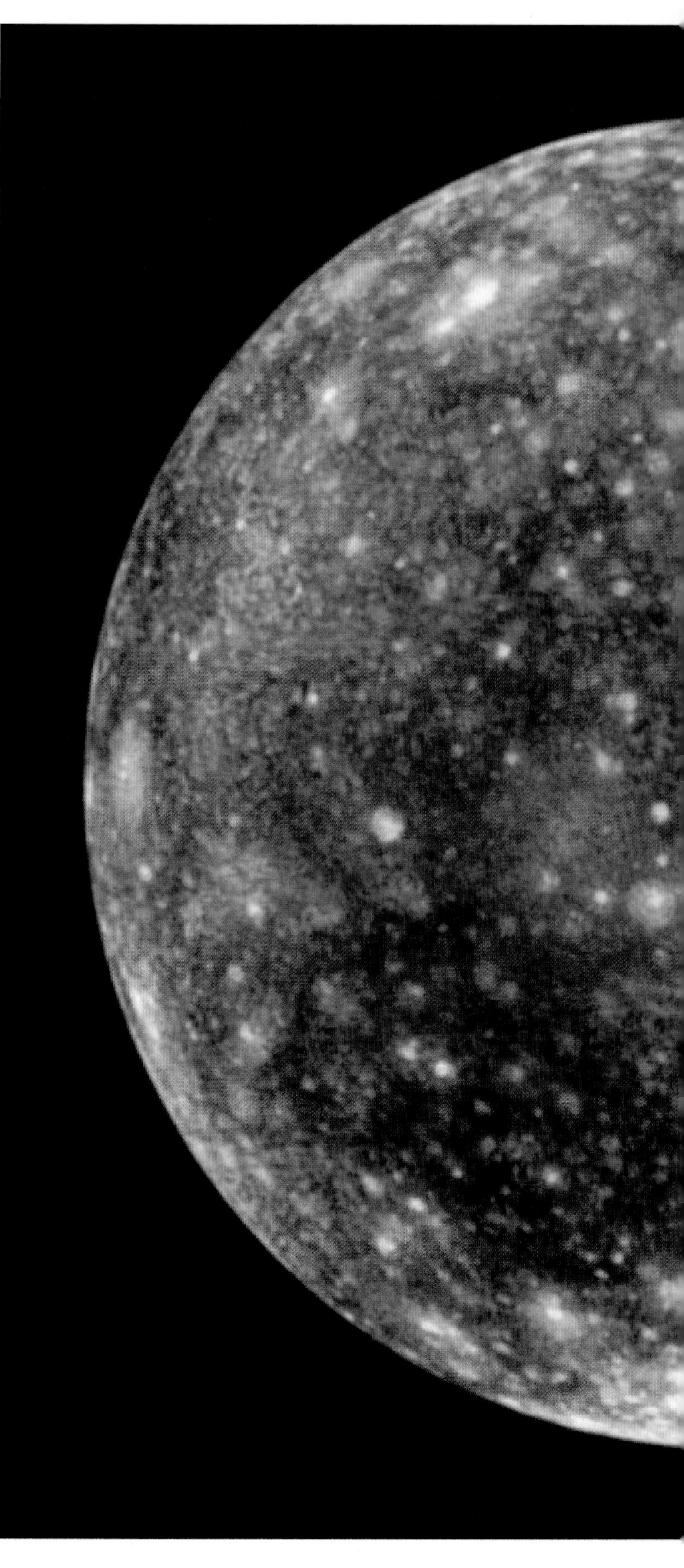

ICE SPIRES ON CALLISTO, VISUALIZATION, 2017

When the Galileo craft descended to just 138 km (86 miles) above the surface of Callisto, it discovered that the planet is dotted with strange towers of ice, 100 m (328 ft) tall. They are covered with dark, dusty material which absorbs sunlight and heats up, slowly melting the ice. The dust slides slowly down the spikes, accumulating at the bottom. Eventually, these will disappear completely. No such landscape has been found anywhere else in the solar system.

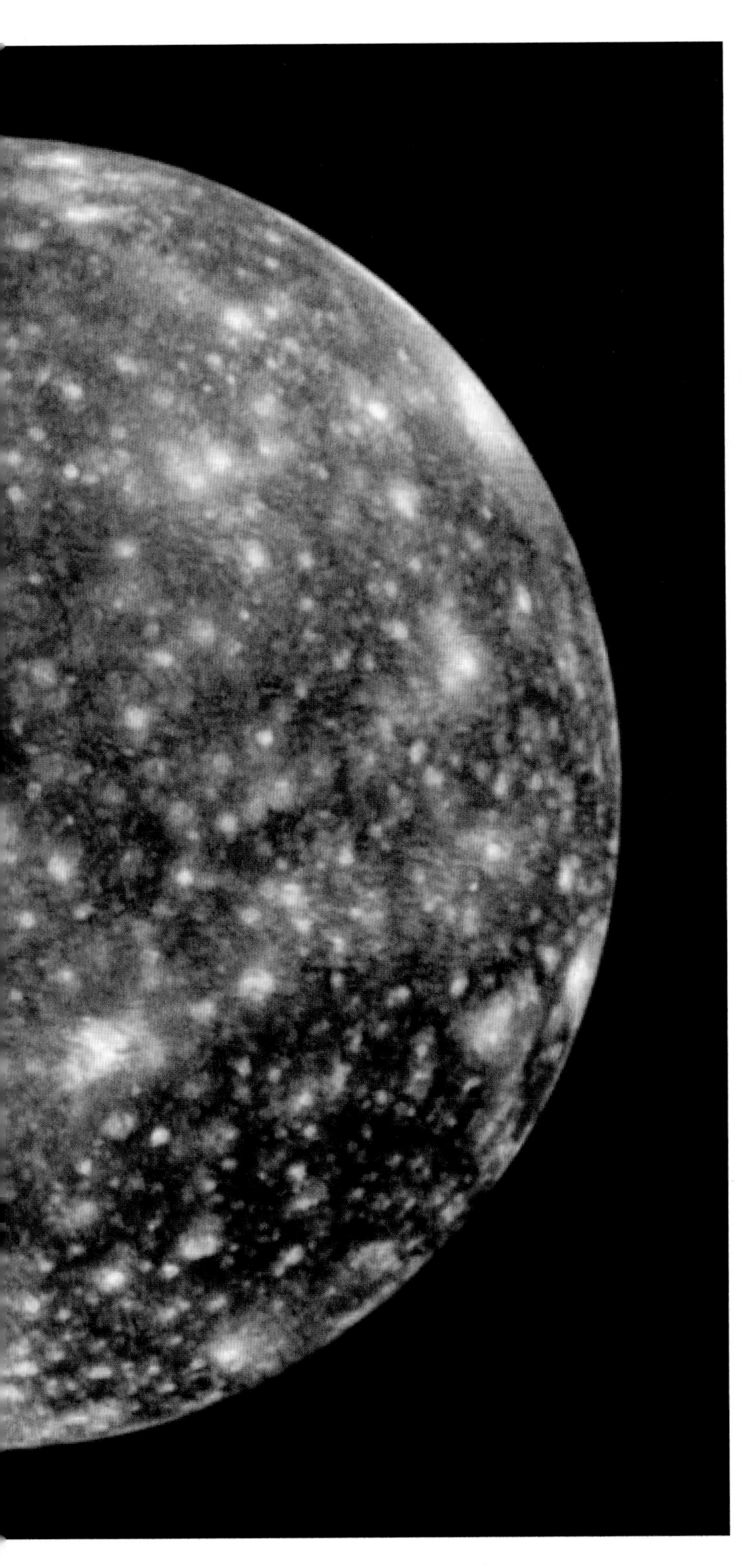

VOYAGER, RINGS OF JUPITER, 2016

Jupiter has a very faint ring system close to the planet. The rings are fed with dust by the four innermost moons. This photo was taken from inside the system, looking out towards the star Betelgeuse.

SATURN

Planetary focus	Saturn
Length of year	29.4 years
Length of day	10 hours
Size x Earth mass	95
Size x Earth radius	9.5
Average distance from Sun	1.4 billion km (869 million miles)
Moons	62
Discovered	Prehistory
Missions	Flyby: Pioneer 11 (1973), Voyager 1 and 2 (1980, 1981) Orbiter: Cassini (2004–2017)

Saturn photographed by Cassini as it passed by the planet in 2004. On the night side of the planet (right), the shadow of Saturn falls across its rings.

Saturn is the second-largest planet in the solar system and the most distant one that can be seen with the naked eye. It has a more extensive ring system than any other planet. It is less dense than water, so in theory would float if dropped into a container of water (if a large enough container could be found).

Like Jupiter, Saturn is a gas giant, made mostly of hydrogen and helium and with no solid surface. It's believed to have a smaller core than Jupiter but the same basic composition, with a layer of liquid hydrogen beneath the thick atmosphere, then a layer of liquid metallic hydrogen, and finally a core of dense metal surrounded by rock. The planet is blanketed by clouds that form stripes, driven by swirling storms and jet streams.

This false-colour image of Saturn was created from data collected by Cassini in 2016 using infrared filters. These filters are sensitive to the scattering and absorption of methane and are useful for revealing the depth and structure of clouds. Adjacent bands of cloud move at different speeds and directions, causing interference and turbulence at their edges.

FIRST VIEWS OF SATURN

The first person to see the rings of Saturn was Galileo in 1610 (see page 97), but he didn't know what they were. They caused him considerable confusion as their appearance through his telescope kept changing and at times they even disappeared.

It was not at all obvious what, if anything, Galileo was seeing near Saturn. At first he suspected there were two large moons, one either side of the planet. But when he looked again two years later, the 'moons' had vanished. The rings of Saturn become invisible from Earth twice every 29.5 years because at two points in Saturn's orbit the rings are edge-on to Earth and too thin to discern, even with a powerful telescope. When the moons returned, Galileo decided that they were 'arms' of some kind, but got no further with his explanation.

The rings of Saturn are very thin. As the planet orbits the Sun, the angle of the rings relative to our view from Earth shifts. Several moons are visible in these photos. In the second one the largest moon, Titan, casts its shadow on Saturn.

The rings disappeared again in 1626, 1642 and 1655. By the later date, there were many more astronomers looking at Saturn through their telescopes. Johannes Hevelius published his explanation in 1656, along with a full set of aspects he believed Saturn could adopt. He claimed that Saturn is elliptical, or oval, with two crescents attached at the ends of the major (long) axis. As the planet rotates around its minor axis, there are times when it is seen edge-on and we see only its width and no crescents. Hevelius' model works in terms of explaining what we see, but would be a pretty odd design for a planet.

Christopher Wren, better known as an architect, offered an explanation closer to the truth in 1658. He suggested that a very thin corona, little more than a film, surrounds the planet and fades to invisibility when the planet is edge-on in our view from Earth. Christiaan Huygens came up with something much closer to the correct explanation in 1659, claiming that the planet is surrounded by a thin flat disk which doesn't touch it at any point; this disk, viewed edge-on, becomes invisible.

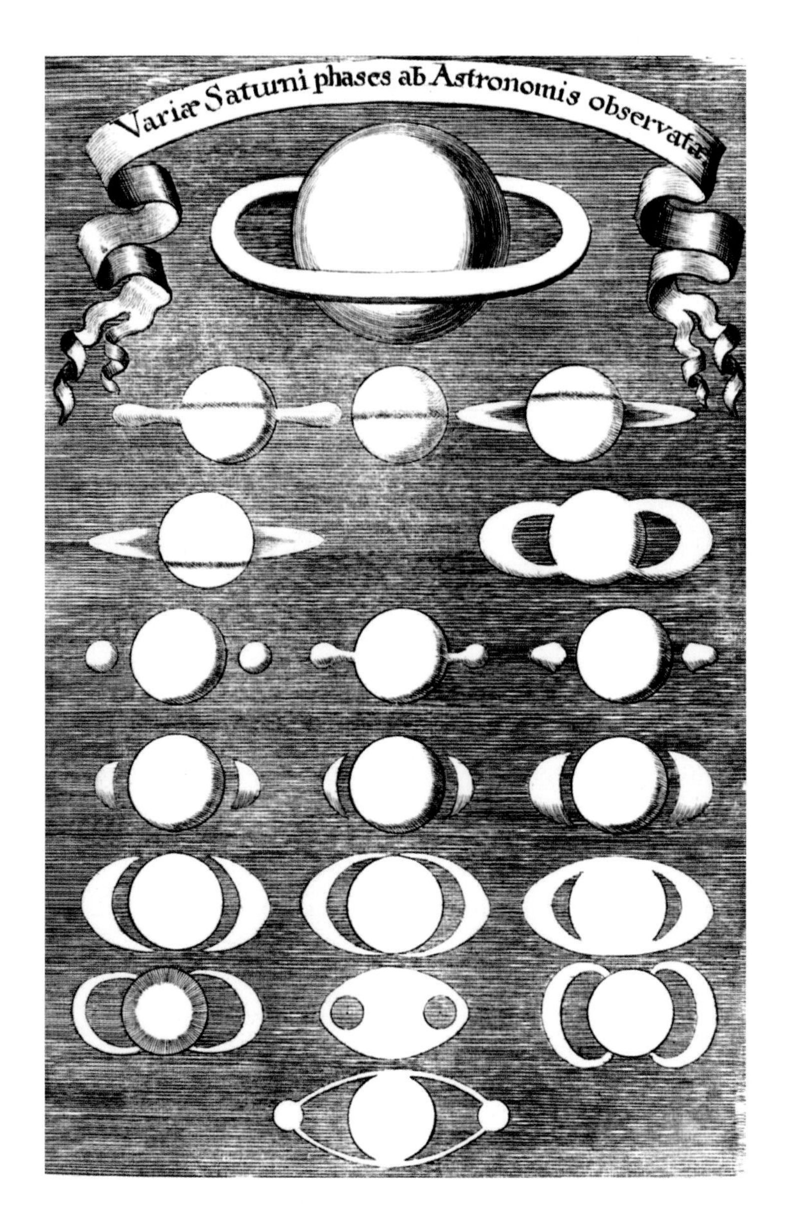

The varying appearance of Saturn, published by Johann Zahn in 1696. The aspect of the planet changes over a cycle of 29 years; its axial tilt means that sometimes the rings are edge-on to Earth and barely visible.

Although Huygens was right about the ring, he was wrong to suppose it is solid. Others at the time suggested it was instead made up of separate pieces of an unknown composition, all in orbit around the planet.

In 1787, Pierre-Simon Laplace demonstrated that a single solid ring could not be stable and suggested instead a nested sequence of discrete – but still individually solid – ringlets. Nearly 200 years after Huygens, in 1858, the physicist James Clerk Maxwell published his account of the rings from a mechanical point of view, demonstrating that they could not be solid, or a continuous fluid, and that the particles making up the rings could each be no more than a few inches across.

Below: *Huygens' solution to the problem of Saturn explains all its various appearances.*

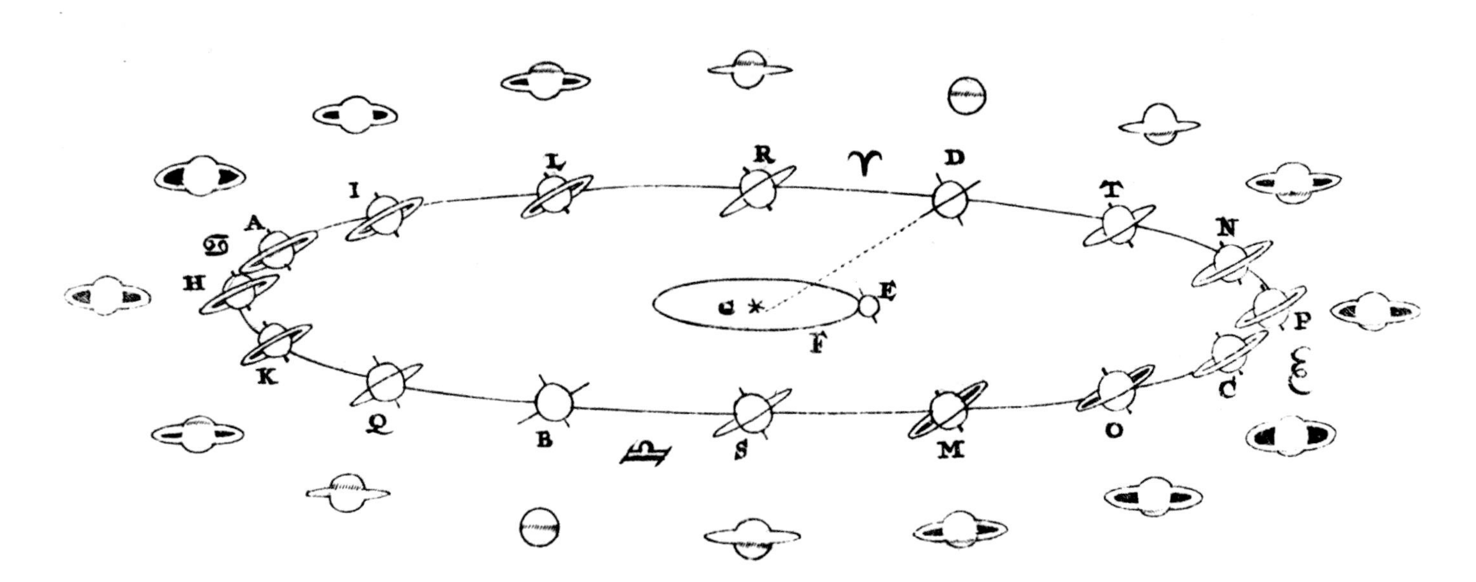

CASSINI, RINGS FROM ABOVE, 2013

This image of Saturn taken from above shows a view that can never be seen from Earth. The rings are fully visible, except where the shadow of the planet falls across them. The clear black line dividing the rings into two areas is the Cassini Division, some 4,700 km (2,920 miles) across.

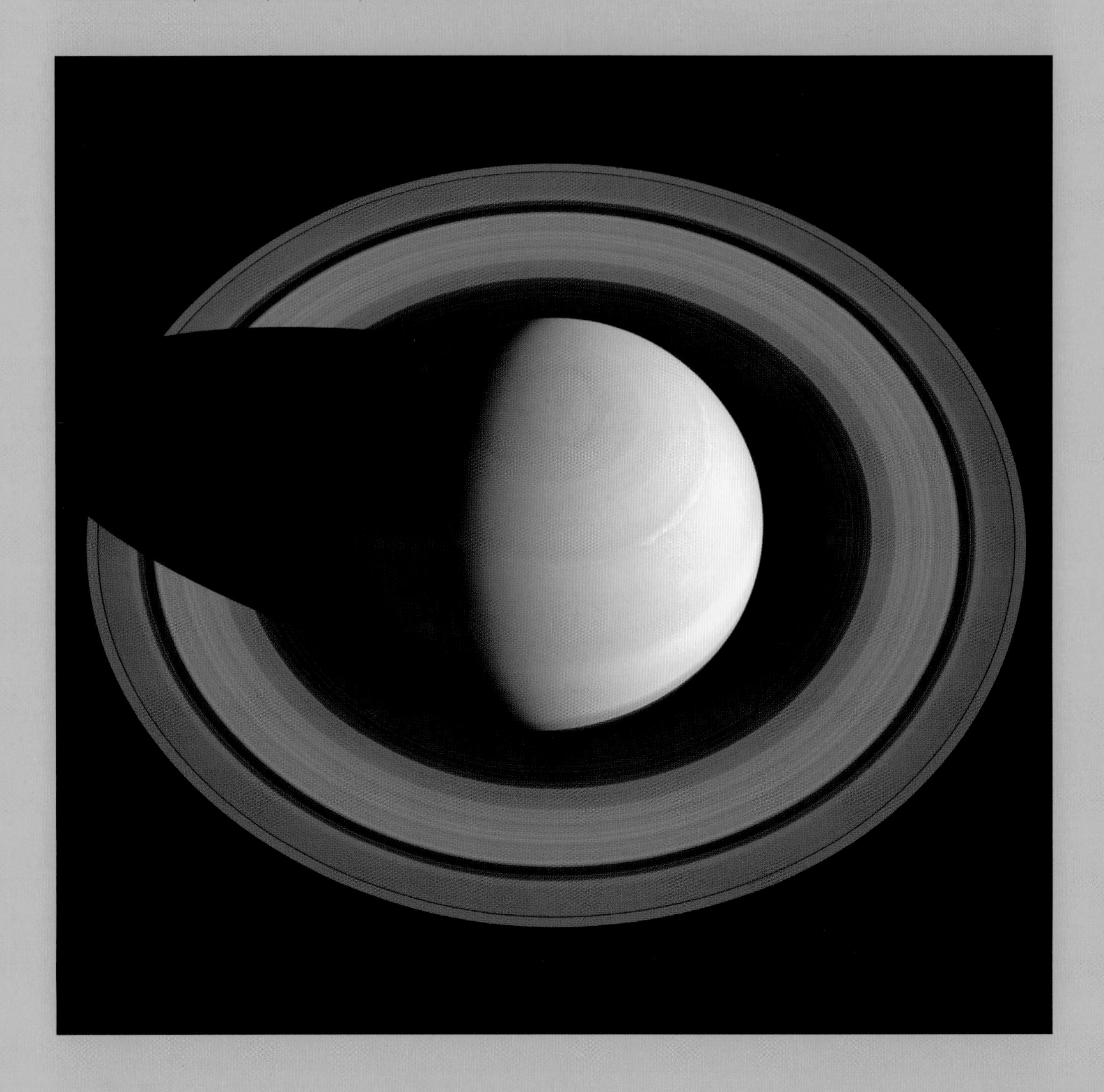

NASA/UNIVERSITY OF COLORADO, MAPPING THE RINGS, 2007

Each of Saturn's rings is made of a huge number of particles of ice mixed with about 0.1 per cent silicate rock or organic compounds, which gives them a reddish tinge. The smallest particles are the size of grains of sand and the largest the size of a mountain; most are in the range 1–10 cm (0.4–4 in).

This false-colour image shows clearly that the rings are not homogenous. The colours indicate the orientation of clumps within the rings, and the brightness shows the density of particles. This density is calculated by measuring the amount of light that penetrates the rings from a star located behind Saturn. In the blue areas, particles clump in tilted 'wakes' produced by gravity drawing them together. In the central yellow regions, particles are too densely packed for any starlight to pass through. The data for the image was collected by the Cassini spacecraft in 2006.

MARTY PETERSON & WILLIAM K. HARTMANN, CLOSE-UP OF RINGS, 1984/2007

This artist's impression of a ring of Saturn in close-up is based on research by the Planetary Science Institute in the 1980s. Icy chunks form loose clumps typically several metres long. The clumps constantly collide, reform and recycle their material. Chunks of ice from the rings rain down on Saturn at the rate of an Olympic-sized swimming pool of water every half hour. The total mass of ice in the system is about half that of the Antarctic ice sheet, but it is spread over an area 80 times the size of Earth.

Data from Cassini suggests that the rings might be a recent development, having formed in the last 10–100 million years. They are eroding slowly and might last only another 100 million years. The short lifespan of the rings (up to 200 million years) suggests that other gas and ice giants might have had – or be about to have – spectacular ring systems that we have missed or will miss.

CASSINI, C AND B RINGS, 2004

This image of the C and B rings (C on the left) is coloured to show how clean or 'dirty' the rings are. The dirtiest rings (those with most rocky dust) are redder. Cleanier, icier rings tend towards the blue end of the spectrum.

CASSINI, B AND C RINGS IN CLOSE-UP, 2009

A Cassini image of the outer C ring and the B ring shows how they are made up of separate smaller 'ringlets' and narrow gaps. Cassini was about 2 million km (1.24 million miles) from the centre of the rings. The rings are mostly about 10 m (30 ft) thick, but in some places matter piles into bumps and ridges up to about 3 km (2 miles) tall.

CASSINI, MOONS AND RINGS, 2011

This single shot by Cassini shows five of Saturn's moons. Janus is on the far left; Pandora orbits just beyond the thin ring in the centre; Enceladus, the brightest, is just above Pandora; Rhea (the largest object in the picture) is cut off at the righthand edge of the image; and Mimas is just to the left of it. The photo was taken at a distance of 1.1 million km (684,000 miles) from Rhea. Moons are important as they provide material for the rings and 'shepherd' them. Moons generally orbit in a clear band within or between rings, as their gravity clears a path.

The Voyager and Pioneer fly-bys of Saturn gave us a first view of some of the planet's moons, but more moons and more detail have been revealed by Cassini. There are still no photos of many of the smaller moons and we know little about them. Some probably formed late or have been captured by Saturn relatively recently. Some moons appear to be still forming. Saturn has only one moon the size of Jupiter's Galilean moons (Titan); it's possible that others formed but have been torn apart by collisions or the gravity of Saturn and their remnants make up some of the smaller moons.

CASSINI, F RING, 2009

Cassini discovered a much larger, near-invisible ring far beyond the rest of Saturn's ring system. It begins at a distance of 6 million km (3.7 million miles) from the planet and extends for a further 12 million km (7.4 million miles). It was discovered in infrared images and is shown here in an artist's impression; Saturn is a tiny dot in the middle.

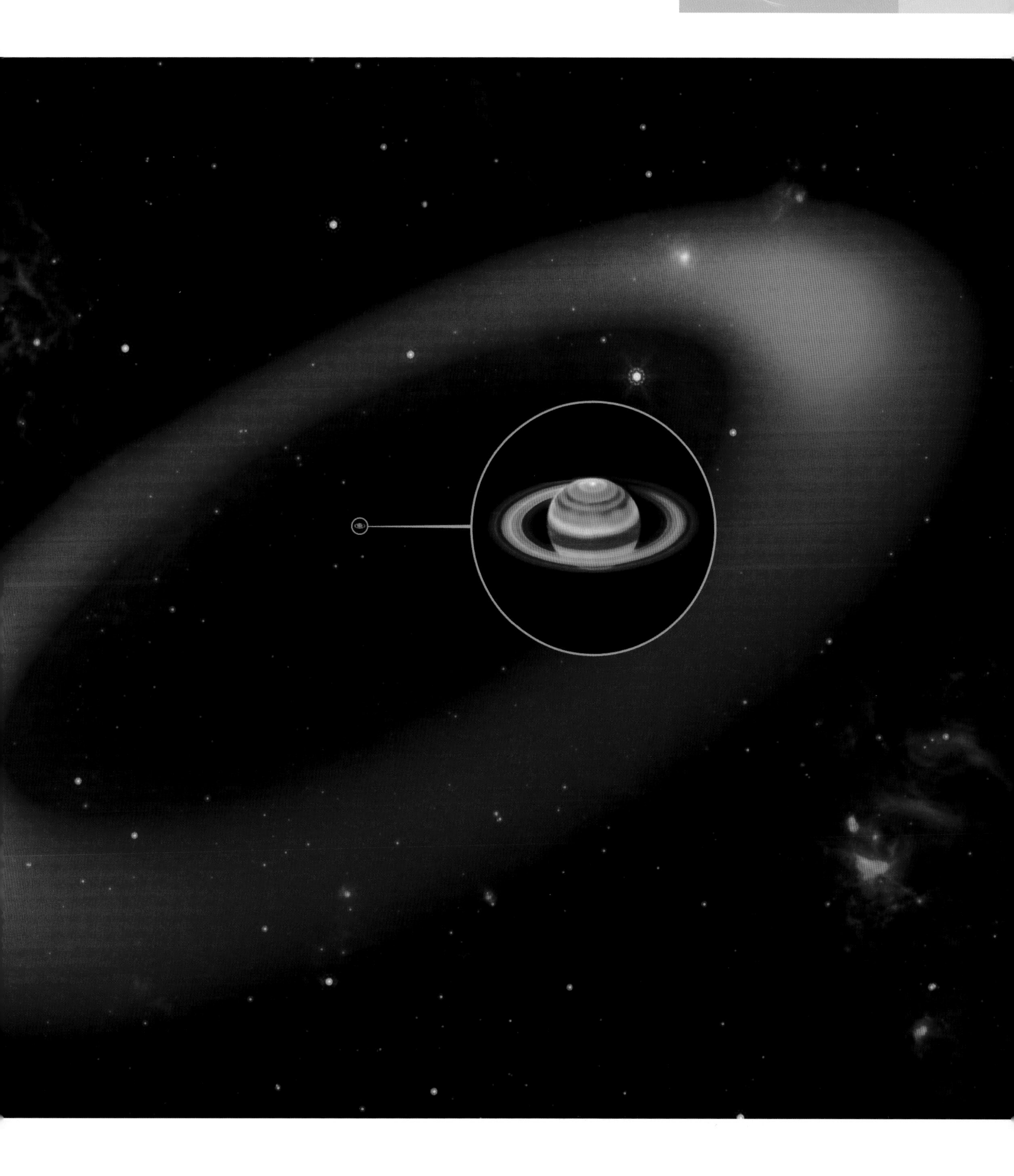

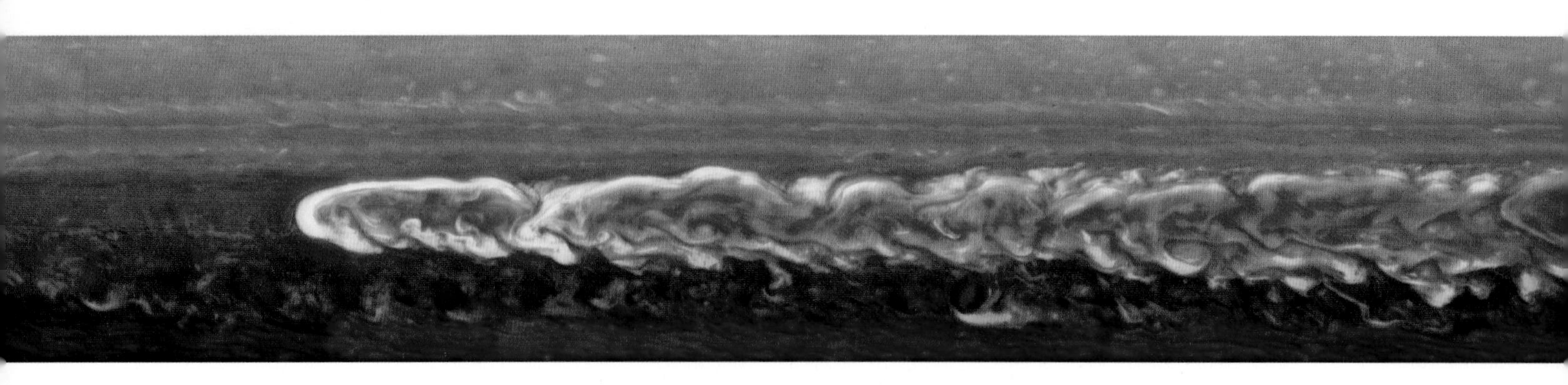

NASA, STORMS ON SATURN, 2011

Blanketed with clouds that circle the planet in jet streams and storms, Saturn is one of the windiest places in the solar system. In the upper atmosphere above the equator, winds can reach speeds of 1,800 kph (1,118 mph). Saturn has three layers of cloud: ammonia at the top, then ammonium hydrosulphide, and finally water. Above the top layer of clouds is a haze of smog. Huge storms, larger than the whole of planet Earth, appear periodically as white areas that break through the clouds. Smaller storms appear as dark spots. The false-colour image above, showing a storm about the size of Europe, is enhanced to make the patterns of the clouds more easily visible. The head of the storm is to the left; a vortex is visible near the centre. The thickest clouds appear white.

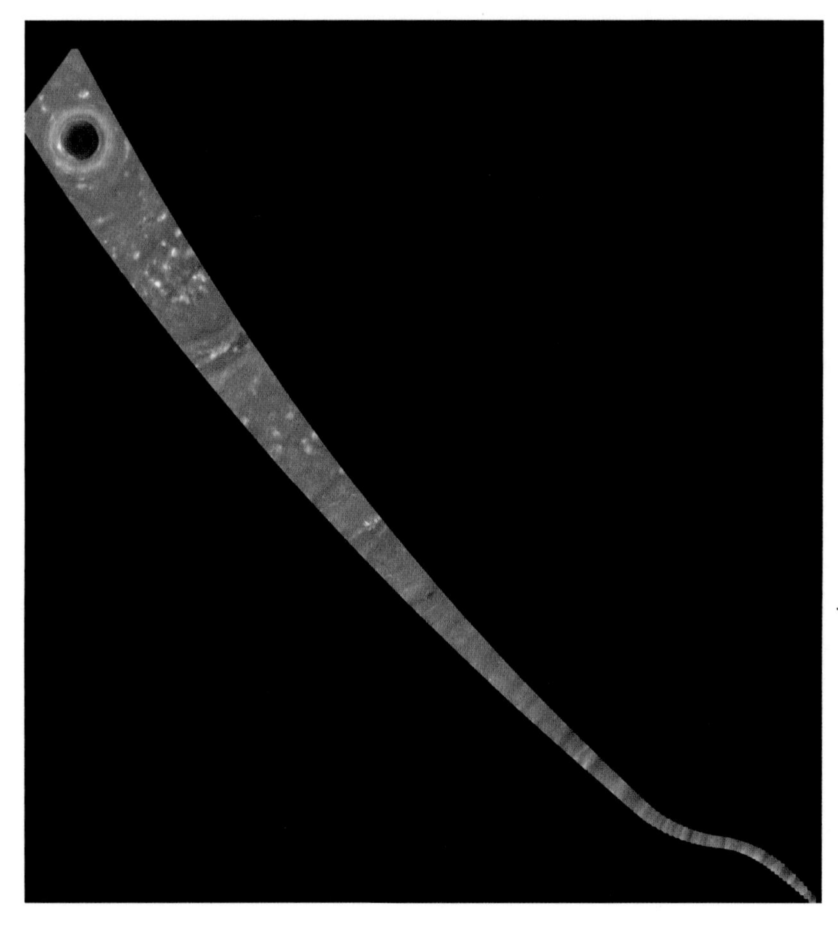

CASSINI, SATURN'S POLAR HEXAGON, 2018

Above Saturn's north pole (facing page), six jet streams converge around a central vortex creating a unique hexagonal storm. The clouds form a tower hundreds of kilometres tall. This false-colour image taken by Cassini shows the structure clearly.

Left: *A mosaic of images from Cassini's final dive towards the planet's atmosphere in 2017, at the end of its mission. The sequence represents acceleration towards the planet, starting above the north polar vortex. At the start of the sequence (top) Cassini was 72,400 km (45,000 miles) above the clouds; by the end, its altitude was 8,374 km (5,200 miles).*

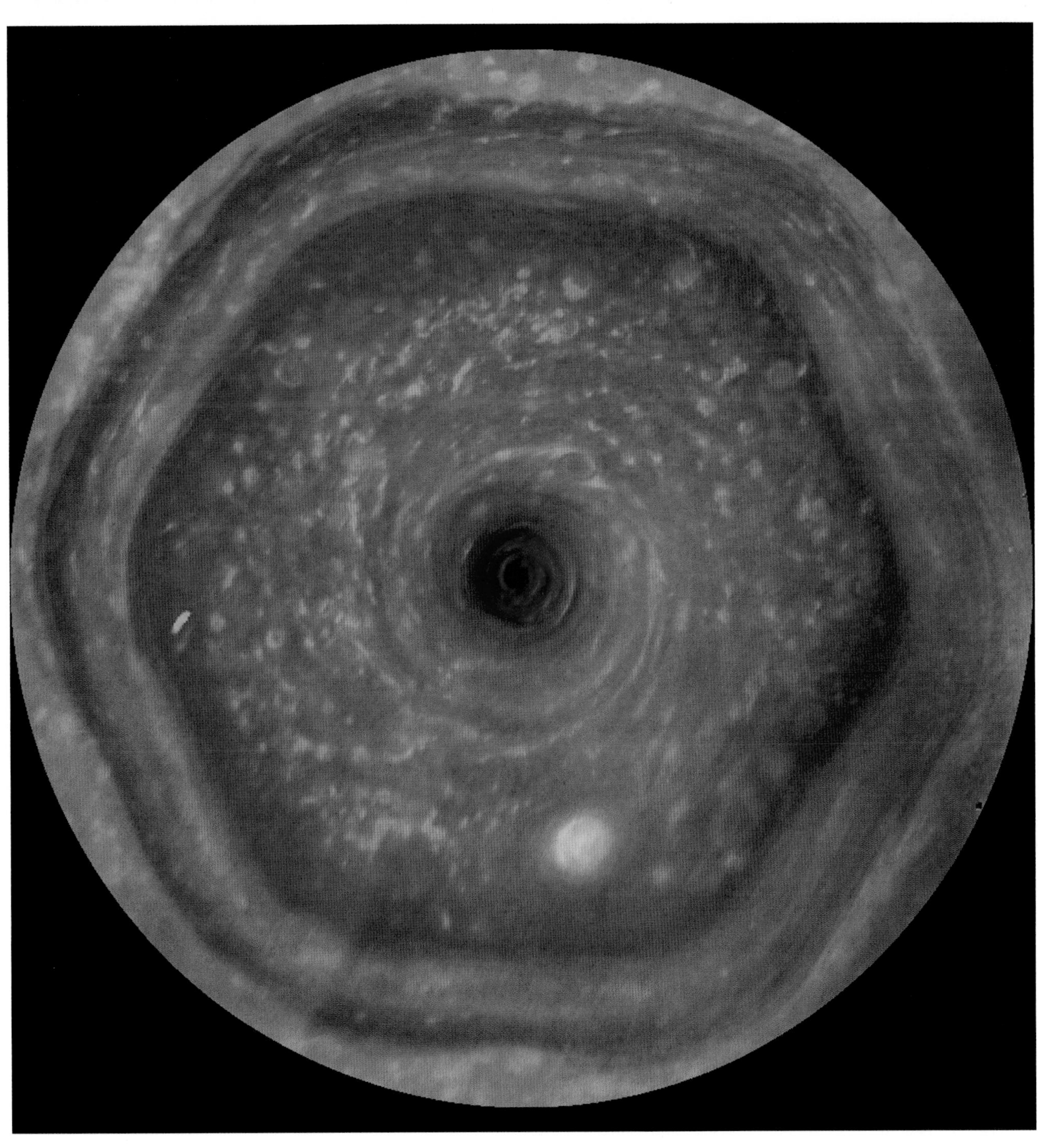

CASSINI, HEXAGON'S SEASONAL CHANGES, 2013/2017

The colour of the hexagon and surrounding clouds changes seasonally, as these true-colour photos taken by Cassini show. The first image is from 2013, the second from 2017. They represent the change that occurs in a bit less than a seventh of a Saturn year. The four-year period is the latter half of Saturn's spring in the northern hemisphere. At this time, ultraviolet radiation from

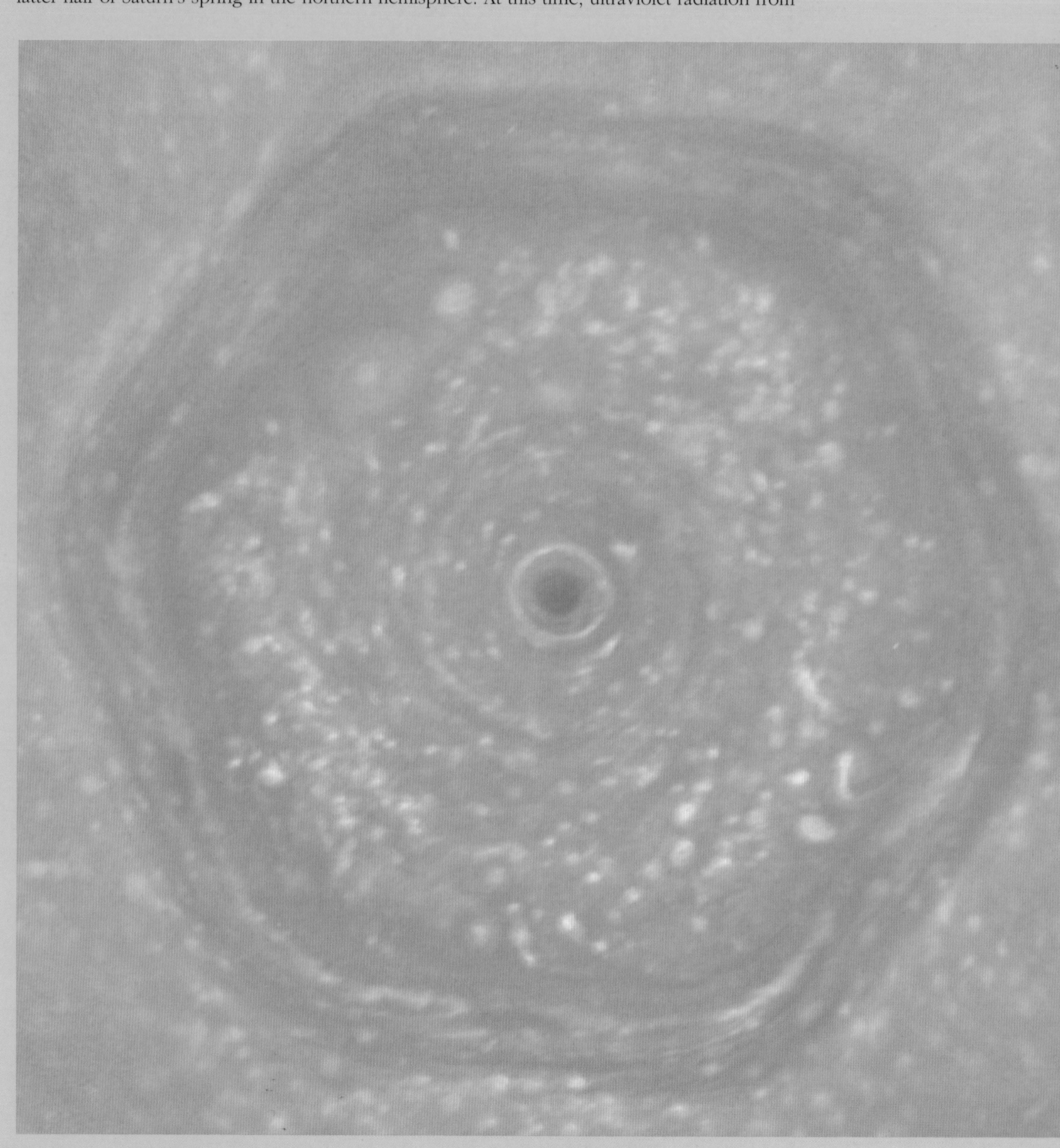

the Sun falling on Saturn's atmosphere starts the formation of photochemical aerosols, which produce a haze of smog above the planet. It's possible that the very centre of the hexagon's vortex is still blue here because the north pole is the last place to be exposed to the Sun's UV rays, and therefore the last part to change. Alternatively, the flow of wind at the centre of the vortex might be downward (as it is in hurricanes on Earth) and this keeps the eye of the storm clear of haze.

THE MOONS OF SATURN

Saturn has at least 62 moons. The first one to be discovered was the largest. Huygens found Titan in 1655, Cassini discovered another four later the same century, and Herschel found two 100 years later. Many of the smaller moons have been discovered by the Voyager and Cassini missions.

Saturn has 24 regular and 38 irregular moons. All the irregular moons are small, ranging in diameter from 4 km (2.5 miles) to 213 km (132 miles).

MOONS AT WORK

Atlas, Daphnis and Pan are among the smaller of Saturn's moons. They are too small to have become spheroid – that requires sufficient mass for gravity to pull the surface inwards with enough force to even it out. All have a job to do: as shepherd moons, they sculpt some of Saturn's rings. Atlas, discovered in 1980, was thought to shepherd the A ring but is now believed only to sculpt a very faint, minor ring. Its thick equatorial bulge has possibly built up by accreting dust from the ring. Tiny Daphnis polices the Keeler gap, a 42-km (26-mile) wide gap in the A ring. Discovered in 2008, Daphnis is 8 km (5 miles) across. Finally Pan, which looks like ravioli, has a sharply defined equatorial ridge. It patrols the Encke gap in the A ring. It is the innermost named moon, and only 35 km x 23 km (22 x 14 miles). It was discovered in 1990.

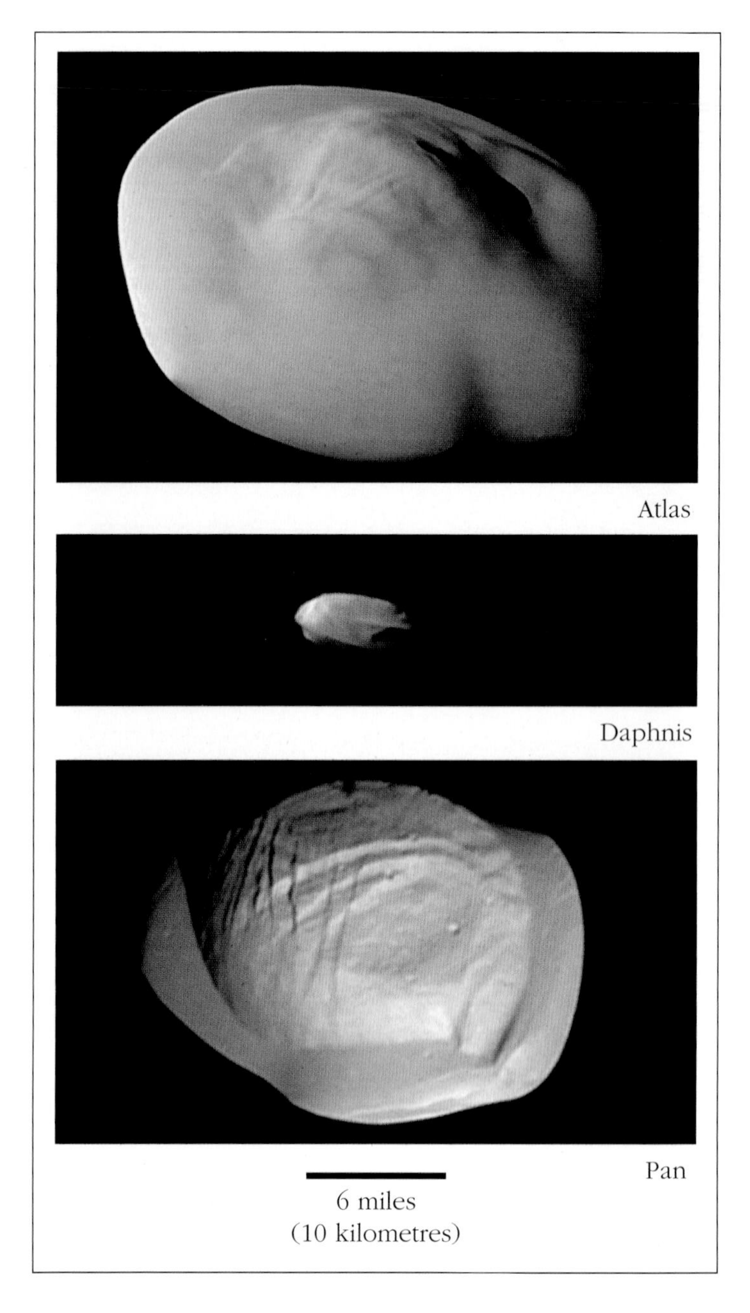

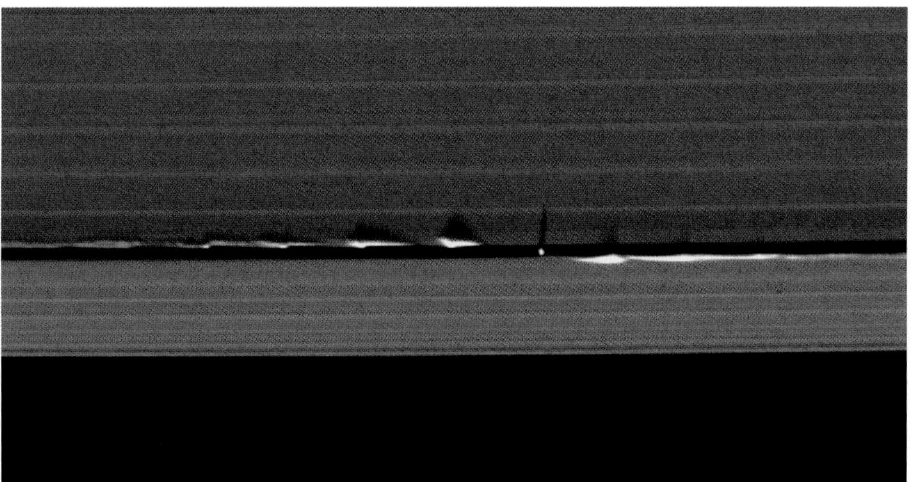

Above: *Three of Saturn's 'shepherd' moons, Atlas, Daphnis and Pan, 2017.*

Left: *A shepherd moon at work. Perturbations on either side of the Keeler gap show the influence of Daphnis, just visible as a bright spot. Particles from the ring are attracted by Daphnis, but not strongly enough to accrete. They fall back into place in the ring as Daphnis passes.*

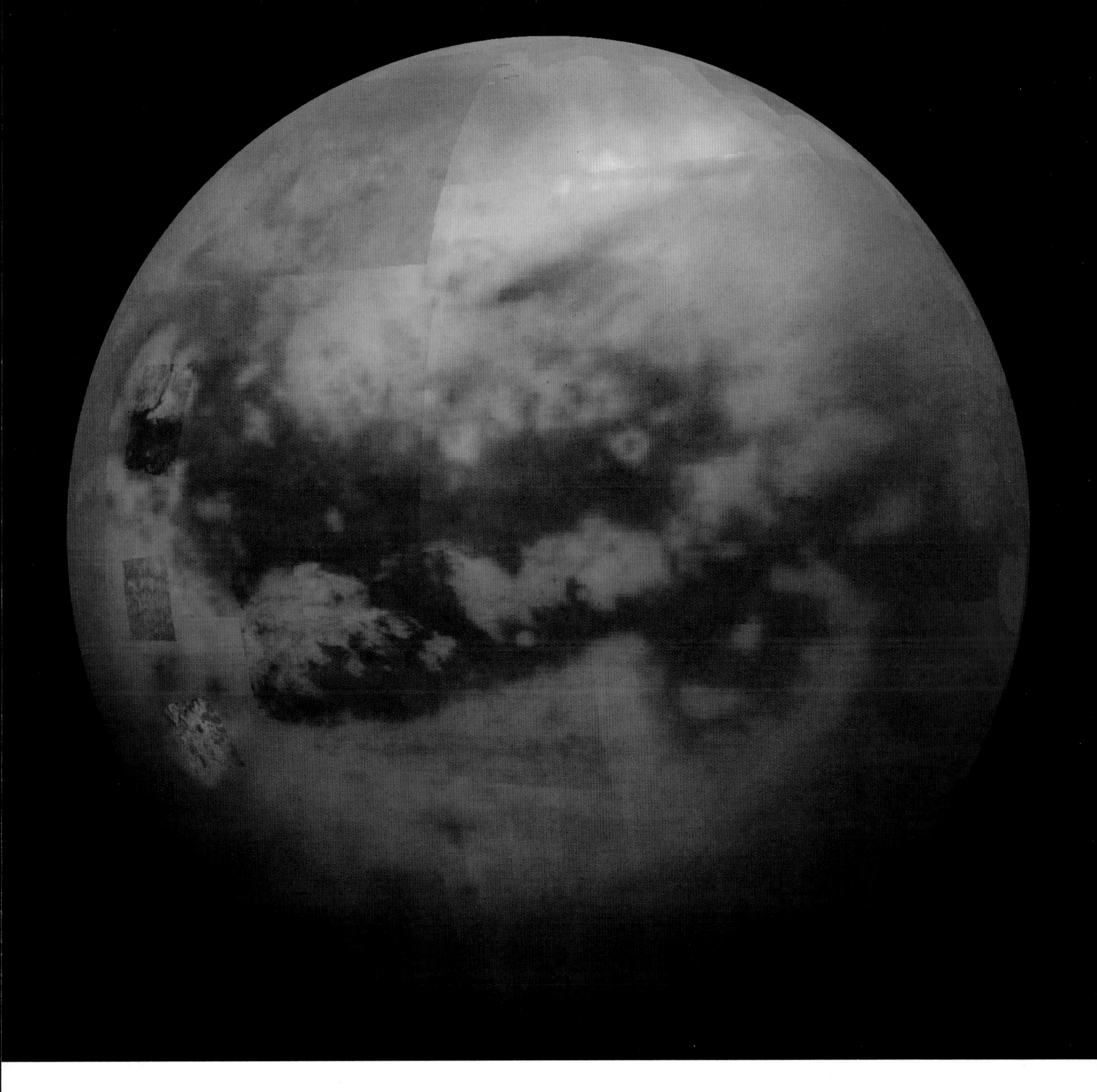

CASSINI, TITAN, 2015

Titan is Saturn's largest moon, bigger than Earth's Moon and larger than the planet Mercury. From space, its surface is hidden behind a dense atmosphere of nitrogen with clouds of methane, as in this photo taken by Cassini. Above that is a thick haze of hydrocarbon compounds – including the (naturally occurring) plastic propylene. Titan is the only moon in the solar system to have a substantial atmosphere. This image shows how infrared pierces the clouds to reveal surface features: the parallel, dark, dune-filled regions called Fensal (north) and Aztlan (south) and Titan's largest impact crater, Menrva (near the left edge, above centre).

ARTIST'S IMPRESSION, DUST STORM ON TITAN, 2018

Titan is thought to have four layers below its atmosphere, with a rocky core surrounded by a layer of high-pressure water ice, then a layer of salty liquid, and a crust of ice. The crust is cloaked with dust composed of solid hydrocarbons that have formed in the cloud layer and fallen to the ground. Periodic dust storms sweep the surface, and dark dunes have accumulated near the equator.

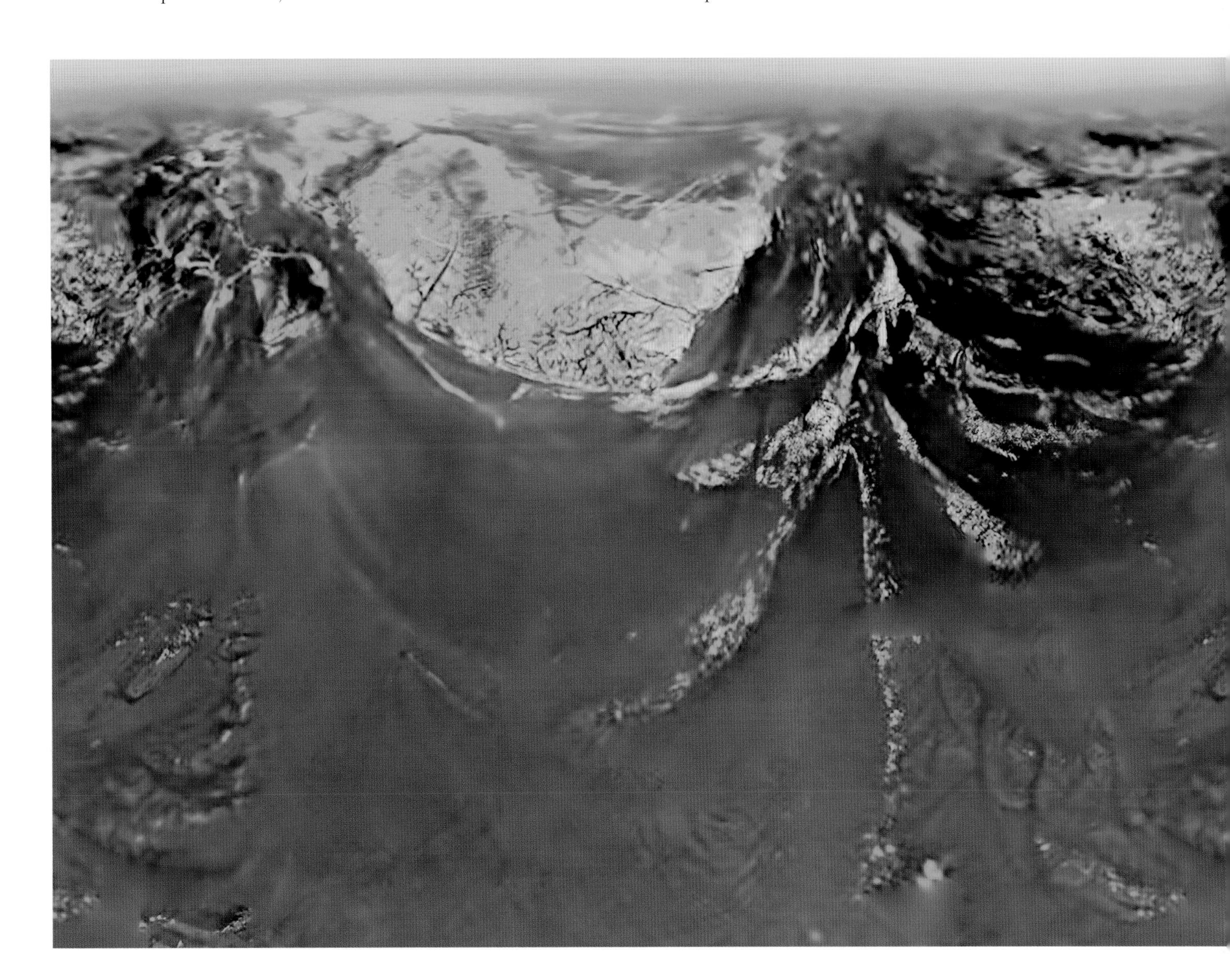

HUYGENS, LANDSCAPE IN CLOSE-UP, 2005

The view of Titan below the clouds has been seen only once, through the images sent back by the ESA Huygens probe dropped by Cassini in 2005.
It returned scenes of a landscape of rugged highlands, deep ravines and the dry relics of once-flowing rivers of hydrocarbons. The surface is rock-hard water ice, and the temperature only -179 °C (-290 °F).

CASSINI, SEAS OF METHANE NEAR TITAN'S NORTH POLE, 2004–2013

Most of Titan's standing liquid is near the north pole, where there are vast seas and scattered lakes. This composite radar image produced from Cassini data shows liquid as blue or black and the terrain as brown/yellow. Liquid is very unevenly distributed over the surface: 97 per cent of the lakes and seas fall into an area 900 x 1,800 km (600 x 1,100 miles). The white areas are unmapped.

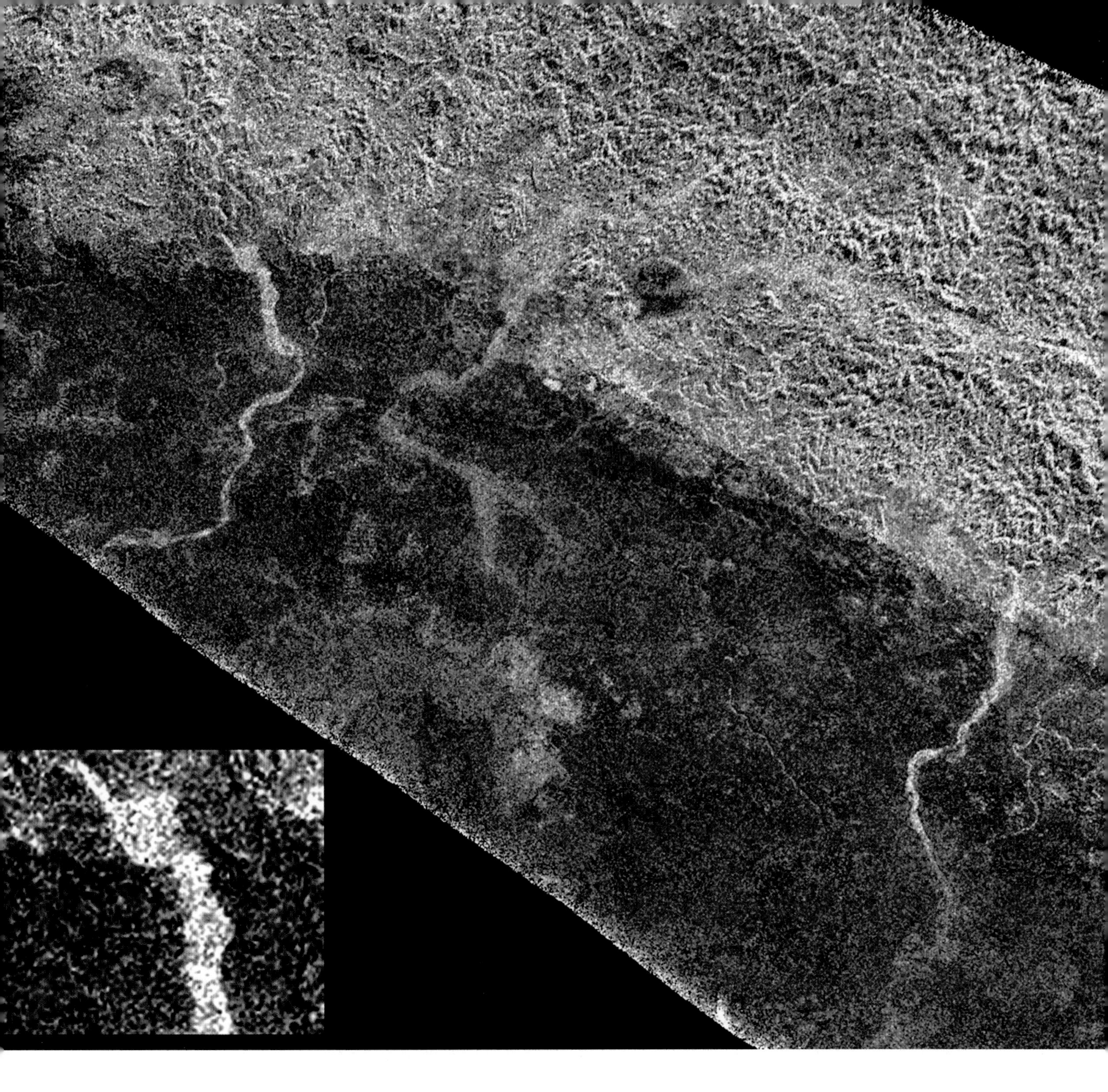

CASSINI, RIVERBEDS ON TITAN, 2008

Titan has rivers, lakes, seas and a cycle whereby evaporated liquid falls as rain. Earth is the only other place in the solar system with a comparable cycle – but whereas Earth has a water cycle, on Titan it is a cycle of liquid methane (natural gas). There might also be volcanic activity, with liquid water from deep below the surface taking the place of lava. The landscape shows the same evidence of running liquid as the terrain on Earth. The sinuous dark lines on this photo taken by Cassini show the path carved out by flowing liquid in an area called Xanadu.

CASSINI, ENCELADUS, 2005

Enceladus is a frozen world, a sheet of ice 30–40 km (19–25 miles) thick, hiding an ocean of salty water that could be 10 km (6 miles) deep. The moon is just 500 km (310 miles) across, the most reflective body in the solar system, and orbits Saturn in just 33 hours.

The distinctive blue 'tiger stripes' are vast cracks in the surface that are 130 km (80 miles) long, 2 km (1.2 miles) wide and 500 m (1,640 ft) deep. They are around 100° C hotter than their surroundings. More tiger stripes have appeared over the time that Cassini has been observing Enceladus. Massive plumes of water ice erupt from them, surging out into space and feeding Saturn's E ring. The erupted ice grains also contain complex hydrocarbons, brought from Enceladus' sub-surface ocean. The presence of water, energy and hydrocarbons makes it possible that Enceladus could support some form of life.

This false-colour mosaic of Enceladus was built up from 21 images taken by Cassini using ultraviolet, infrared and visible light.

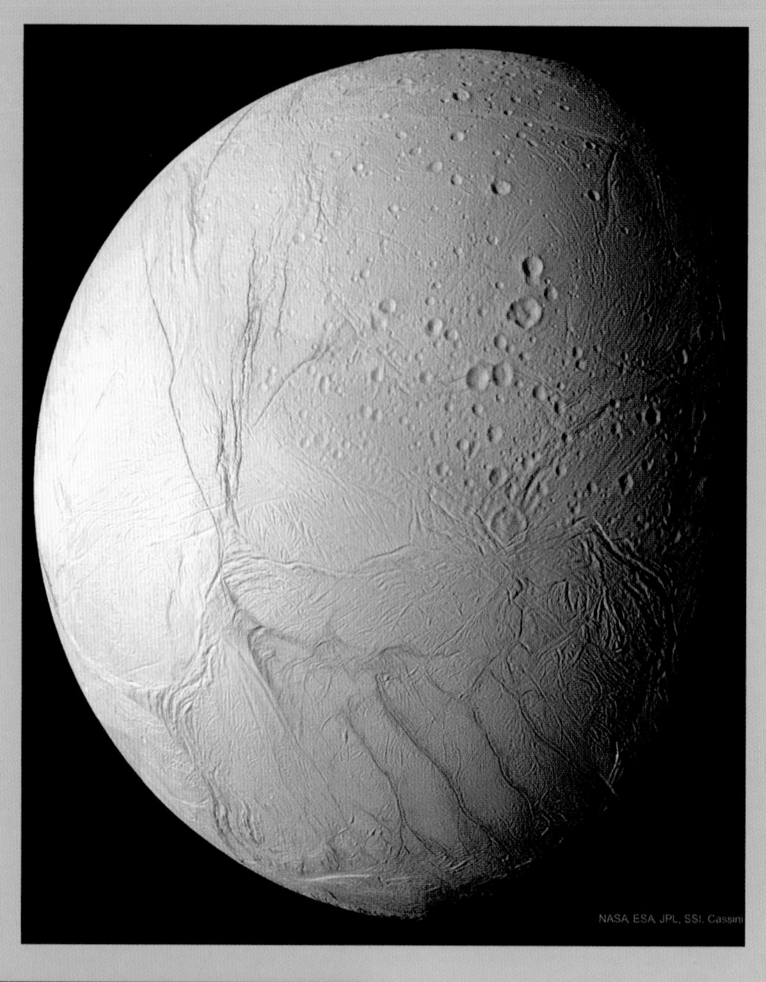
NASA, ESA, JPL, SSI, Cassini

Plumes of water ice erupt into space around Enceladus at 1,290 kph (800 mph).

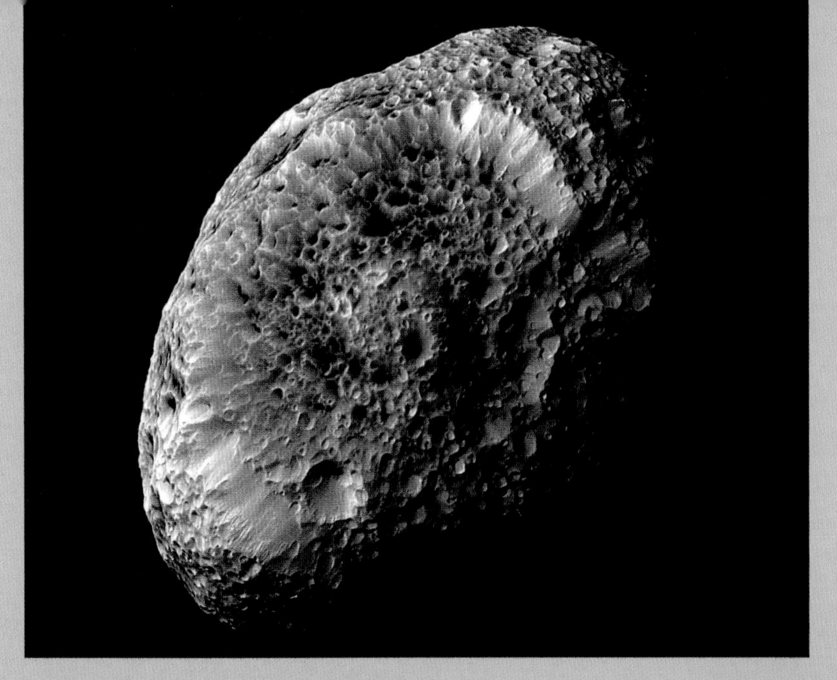

CASSINI, HYPERION, 2005

Hyperion, discovered in 1848, looks like a giant potato-shaped sponge and is the largest non-spherical moon in the solar system. Deeply cratered, it has a density so low that it might house a huge system of caverns inside.

CASSINI, MIMAS, 2017

Mimas, discovered by William Herschel in 1789, has one of the most cratered surfaces in the solar system. This moon is composed mainly of water ice with a small amount of rock. The craters on Mimas record its entire history, as nothing has erased or weathered them. The biggest crater and most noticeable feature is the Herschel crater, 130 km (80 miles) across – a third of the diameter of the planet. The walls of the crater are 5 km (3 miles) high and it has a central peak slightly taller at 6 km (3.5 miles). Cracks called 'chasmata' on the other side of the moon were possibly created by the shock waves of the impact that made the crater, and must have come close to blasting the moon apart. Mimas is only 186,000 km (115,000 miles) from Saturn and orbits the planet in just 22.5 hours.

CASSINI, IAPETUS, 2007

Saturn's third largest moon is a world of striking contrasts. One side is as dark as coal, with a red tinge, and the other is bright white. Evidence from the Cassini flyby in September 2007 suggests that this could have been caused by a combination of incoming dust, followed by ice migrating from the warmer (dark) side to the colder (bright) side. Iapetus is thought to be about three-quarters water ice and one-quarter rock.

FROZEN WASTES

GLOBES OF HOT SLUSH WITH AN ICY CRUST

The ice giants Neptune and Uranus lie beyond the gas giants. The name 'ice giant' suggests a frigid world of solidified water, but that's far from the case. The planets are made of a mix of water, ammonia and methane, their interiors are slushy rather than solid, and 'ice' can be hot as well as cold. They are so far from Earth that they were not known to the ancients as planets. Neptune is not visible to the naked eye, and Uranus is so distant that its movement across the sky is very slow and it was not recognized as a planet.

The two most remote planets, Uranus (left) and Neptune (right), photographed (separately) by their only visitor, Voyager 2.

PLACES OF MYSTERY

We know far less about Uranus and Neptune than the other planets. They have been investigated by telescope since the 18th century, but visited only once, on a flyby by Voyager 2 more than 30 years ago. Neptune and Uranus each have an atmosphere of 83 per cent hydrogen, 15 per cent helium and 2 per cent methane and other hydrocarbons.

Around 68 per cent of their mass is made up of ice – but this is not always cold. Within the ice giants, liquid water, ammonia and methane are compressed under such great pressure that they form a slushy, hot ice.

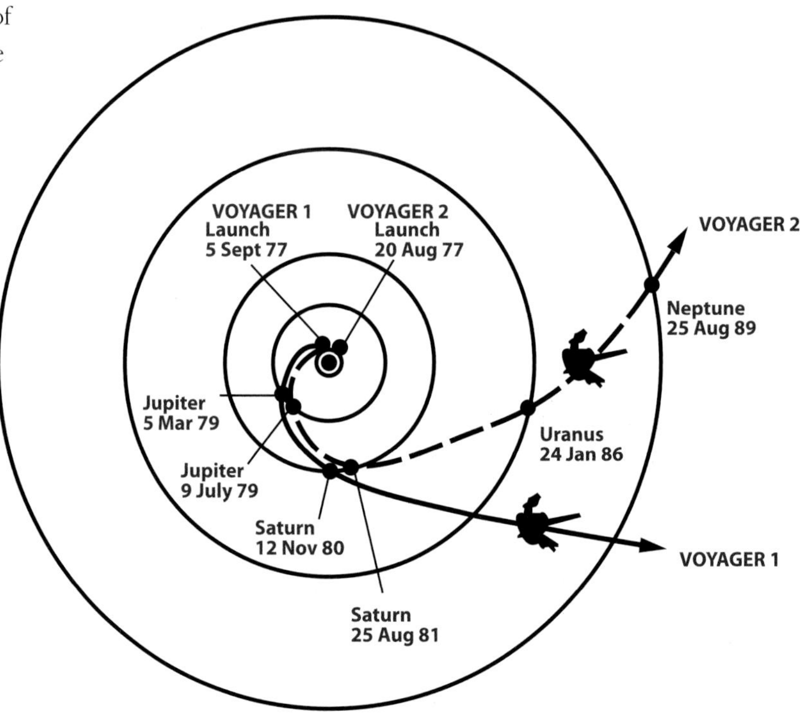

The pair of Voyager spacecraft, 1 and 2, were launched in 1977. After flying past the outer planets, they are now heading off into interstellar space.

The likely internal structure of an ice giant. Beneath the atmosphere is a largely liquid mantle, perhaps slushy further down, then an outer core of ice, and finally an inner core of at least semi-liquid molten metal and rock.

DIAMOND RAIN

The dynamics inside an ice giant are not known for certain. It's thought that the water and ammonia may exist as electrically charged liquids, but that the methane condenses, splitting into carbon and hydrogen. The carbon would then crystallize, forming diamonds which might rain (or hail) down on the inner part of the planet. Simulating the conditions within Neptune in a laboratory in 2017 – temperatures of 2,000–3,000 Kelvin and 100,000–500,000 times atmospheric pressure – scientists produced a mix of solid diamonds in liquid hydrocarbons. If diamond rain falls on Neptune, it doesn't fall near the nominal surface but 7,000 km (4,350 miles) below.

There is a crucial difference between Uranus and Neptune that can't be explained with confidence. Neptune generates ten times more heat energy than Uranus, and makes more heat than it gains from the Sun. It's possible that diamond rain fuels Neptune. Falling diamonds near the planet's core could convert gravitational potential energy to heat.

FUTURE MAPPING OPPORTUNITIES

Both ice giants are a very long way away: they orbit billions of kilometres from the Sun. A visit to either would have to be timed carefully to coincide with the planet's closest approach to Earth's orbit, and as they both have orbits of over 100 years the opportunities are scarce. By good fortune, the next opportune time to launch a mission to Neptune is 2029–2034. After launch, it will still take 10–13 years for a spacecraft to reach the planet. A mission has been proposed, but not yet approved.

URANUS

Uranus is unlike any other planet in the solar system in that it lies on its side at an angle of 98° to its equator. The rings circle the planet almost vertically in relation to the plane of its orbit around the Sun, and its poles are on a near-horizontal axis. It is believed that the planet's odd position was caused by a long-ago collision that knocked it over. Venus, the only planet that

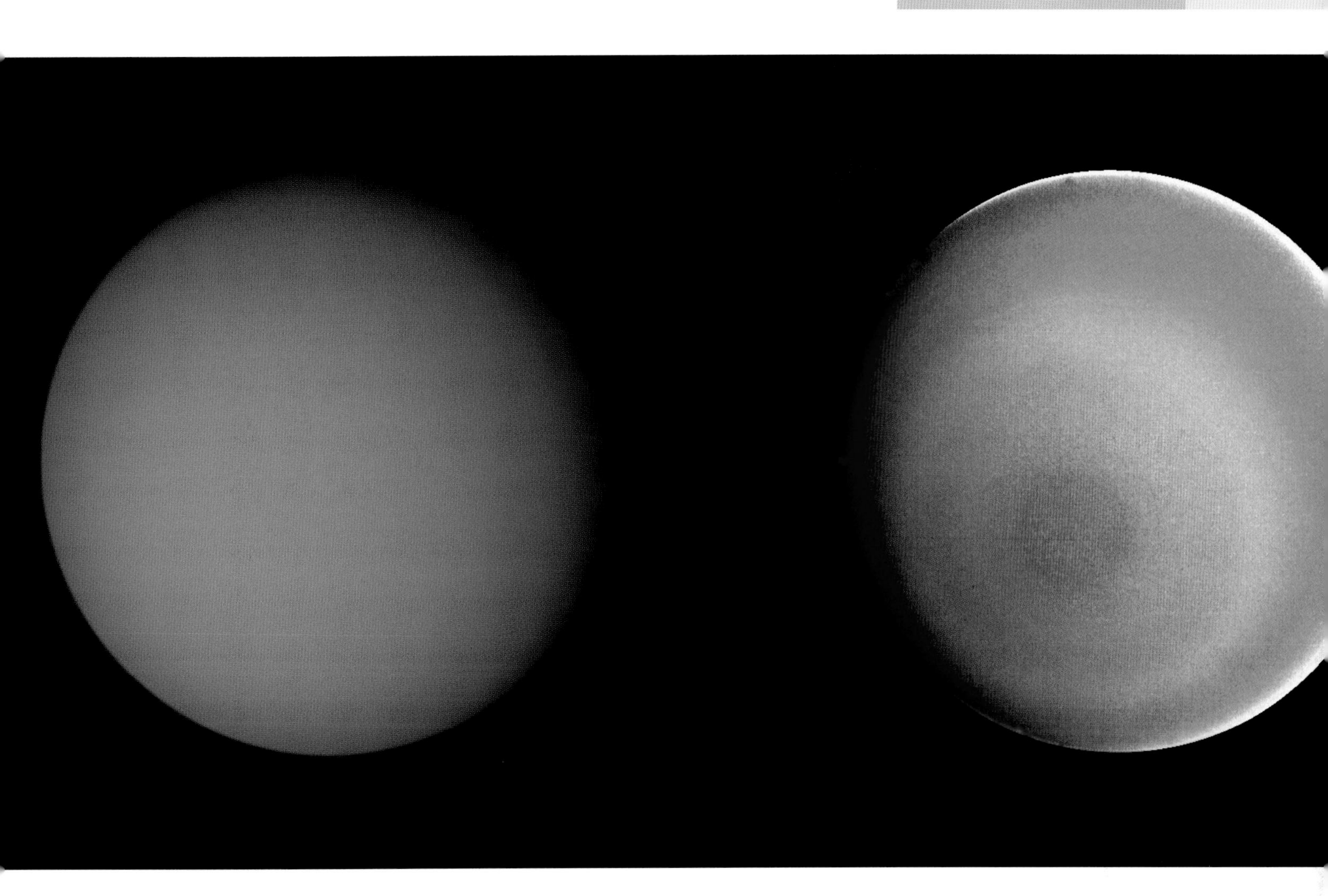

Planetary focus	Uranus
Length of year	84 years
Length of day	17 hours
Size x Earth mass	14.5
Size x Earth radius	4
Average distance from Sun	2.9 billion km (1.8 billion miles)
Moons	27
Discovered	1781, William Herschel
Missions	Flyby: Voyager 2 (1986)

spins 'backwards' (east to west), is also thought to have been the victim of collision. The axial tilt of Uranus means that while the equator has a normal day/night cycle, the poles experience a 42-year day and a 42-year night.

Computer simulations suggest the impactor that knocked Uranus over was a rocky planet about twice the size of Earth and that it hit Uranus a glancing blow. Material blasted out and the impactor would have fallen into the planet, leaving some debris in orbit to form the moons and a faint ring system. Lumps of displaced rock within the mantle might explain Uranus' irregular magnetic field. The impact was not severe enough to blast away the atmosphere of Uranus or alter its orbit around the Sun.

The false-colour image of Uranus (above right) reveals slight differences in the planet which looks uniform in natural colour (left). The nature of the differences is not entirely clear, but one theory suggests that a brown-tinged haze concentrated over the pole is arranged into bands moving in the upper atmosphere.

THE FIRST NEW PLANET

Uranus was discovered in 1781 by amateur astronomer William Herschel. Building his own high-quality telescopes, Herschel set out to search for double stars; in 1779 he began to look at every bright star he could see and, in 1781, found one that moved relative to the background stars. Its motion meant it was not a star but an object within the solar system. Assuming it to be a comet, Herschel notified professional astronomers, who quickly plotted its orbit. Herschel's 'comet' was clearly a planet – one very far from the Sun and never noticed before.

The first planet to be discovered through the use of a telescope, Uranus made Herschel an instant celebrity. He proposed to name it after King George of England, but his suggestion was vetoed and, like its neighbours in the solar system, the planet was given a name from classical mythology. Even so, the King gave Herschel a generous pension which enabled him to build more telescopes, give up his day job, and become a full-time astronomer.

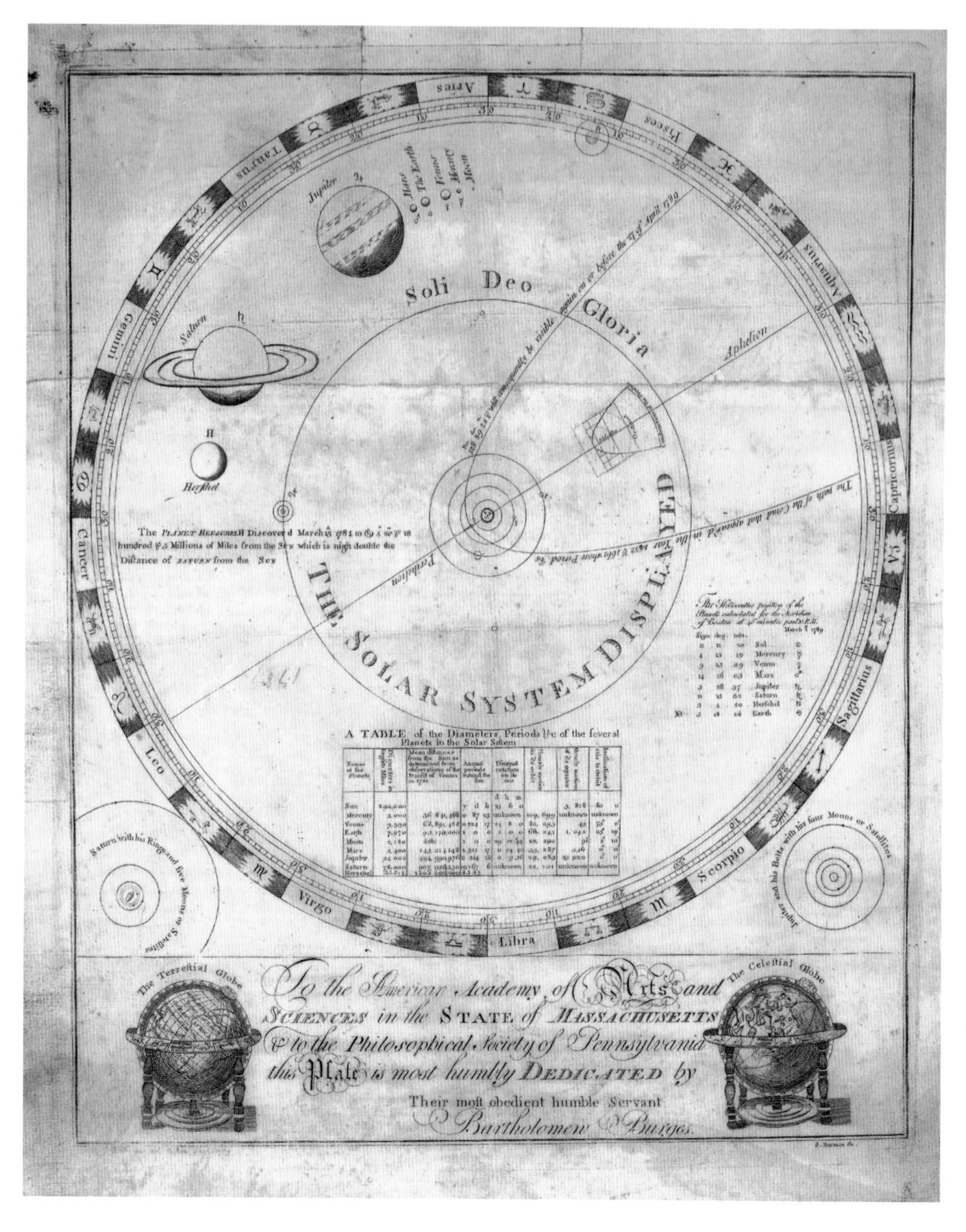

This map of the solar system published in 1789 shows Saturn, Jupiter and Uranus (here labelled 'Herschel') in a band well outside the orbit of the inner planets, but not given distinct orbits of their own.

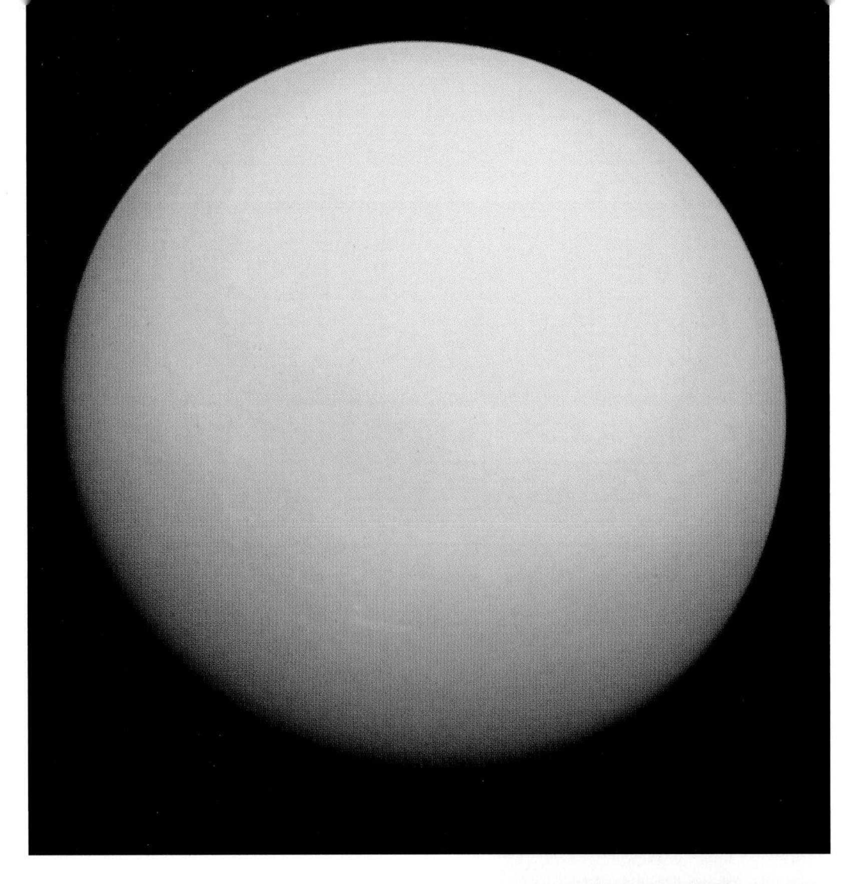

EASILY MISTAKEN

Uranus' orbital time of 84 years means it moves very slowly against the background stars, so earlier astronomers assumed it was stationary. It was first recorded as a star by Hipparchus in 128 BC.

Left: *Uranus photographed by Voyager 2 in 1986 (in natural colour) appears an entirely featureless world.*

BLUE SKY WITH CLOUDS

Uranus looks blue from the outside because the methane in its atmosphere reflects blue light. Thick banks of cloud and storms are not as prevalent here as on the gas giants, and Uranus has a more uniform appearance because its clouds are deep within the atmosphere. The very top of the atmosphere also collects hydrogen sulphide – so if we were able to visit Uranus, it would smell like rotten eggs.

As with the gas giants, parts of the atmosphere move more quickly around Uranus than the planet itself rotates. At some latitudes, winds are so fast that a storm can circle the entire planet in less than a day.

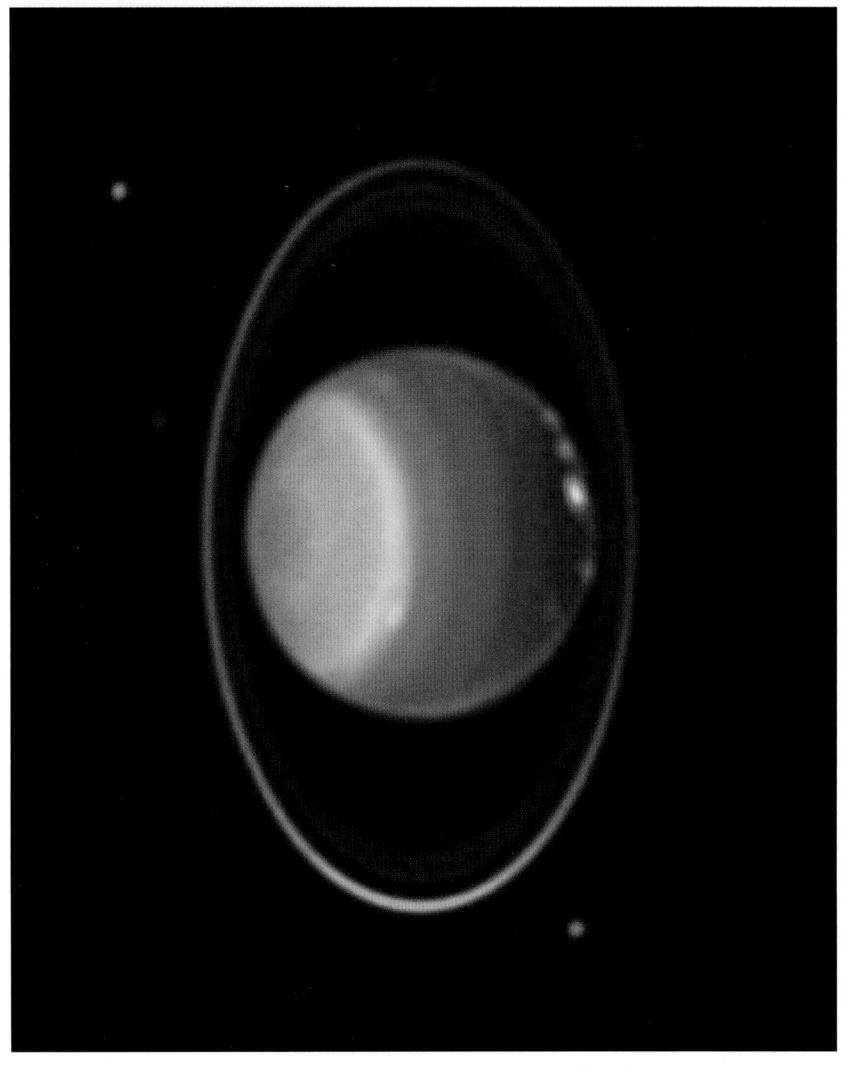

A Hubble Space Telescope image from 2000 shows Uranus' four main rings and ten of its moons. The false-colour image reveals bands of cloud beneath the uniform top layer of the atmosphere.

HUBBLE, CHANGING RINGS, 2003–2007

Five rings were found around Uranus in 1977, four more soon after, then another two were discovered by Voyager in 1986. In 2005, the Hubble Telescope revealed the two outer rings, which are very faint and much further from the planet than the inner rings. Since the discovery of the rings, there has not yet been time to observe them for a full orbit of Uranus. In 1797, Herschel said he believed he could see a reddish ring, but it was not mentioned again by him or any other astronomer. The rings are considered too faint in their current state for Herschel to have spotted any of them, though his description does match one of the outer rings.

The images on the right show how the appearance of the rings has changed with time. The glare from Uranus has been blocked out, and the planet has been added back in afterwards to show its position and size. As a result, the rings don't run across the face of the planet.

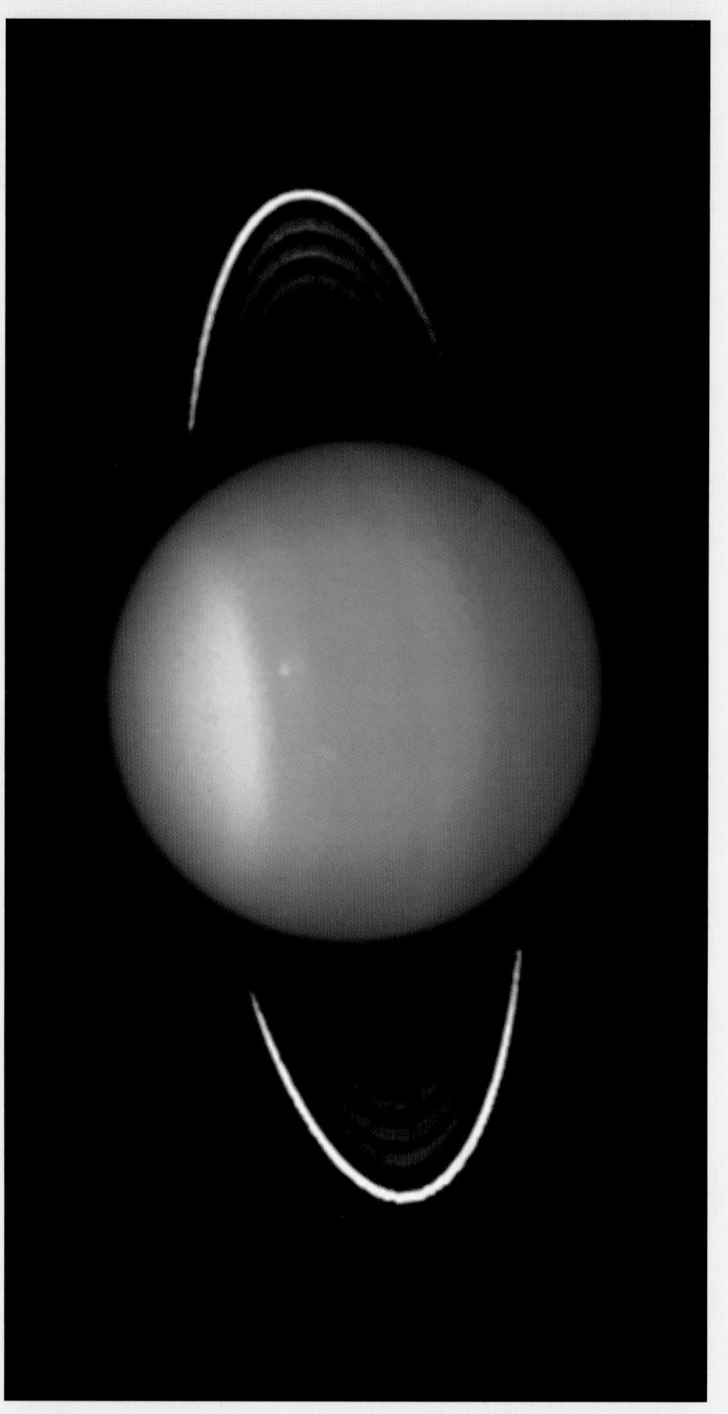

2003

VOYAGER 2, RINGS IN FALSE COLOUR, 1986

This false-colour image of the inner rings of Uranus makes them clearly distinguishable. Unlike the rings of the gas giants, Uranus' rings are made up of very small and dark particles rather than ice. The size of particles in the brightest ring is

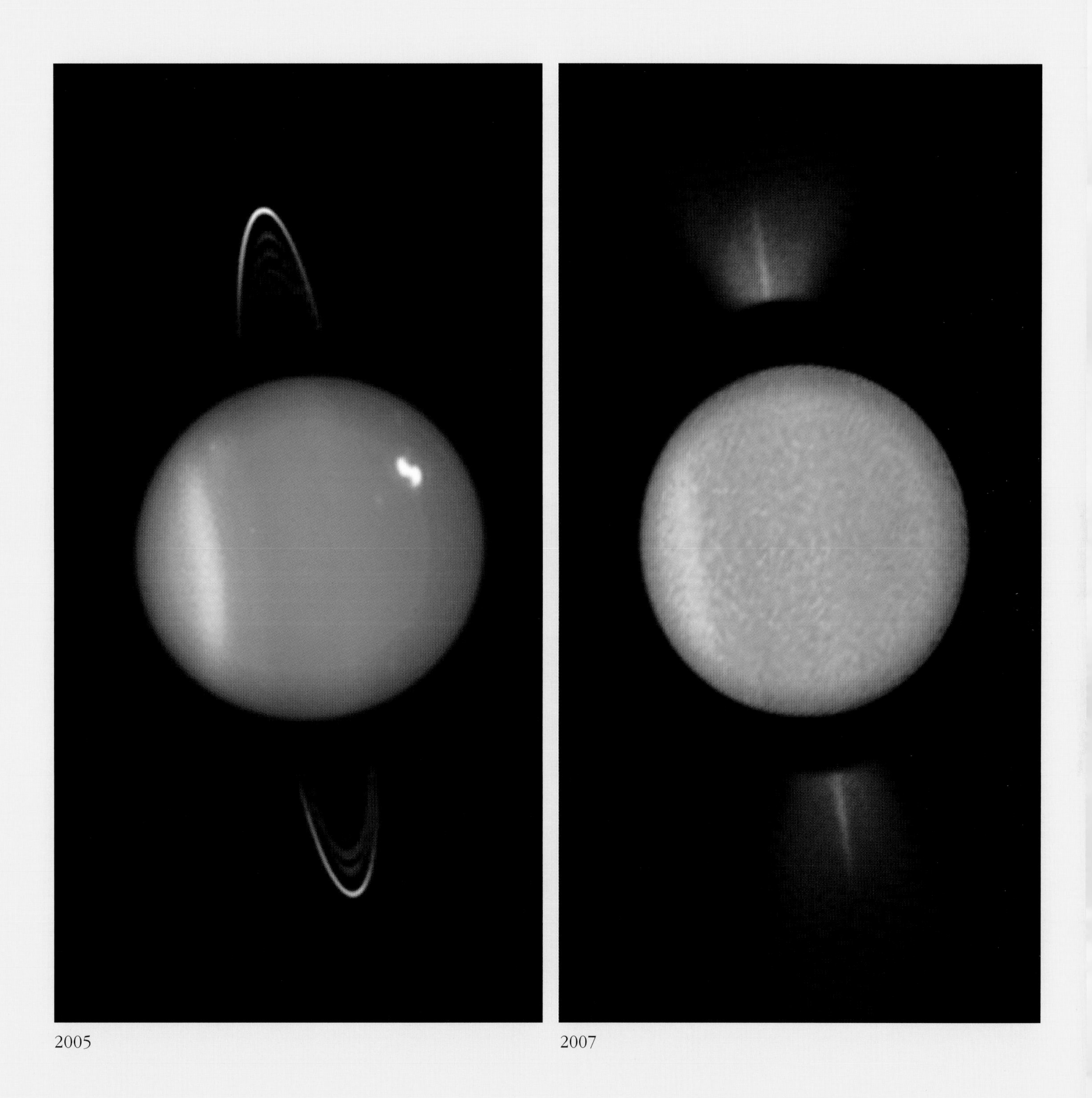
2005 2007

0.2–20 m (8 in–65 ft), and the ring is only 150 m (492 ft) thick. The photo was taken from a distance of 4.2 million km (2.6 million miles) and shows, from left, the rings epsilon, delta, gamma, eta, beta, alpha, 4, and 6. The space between rings is not clear, but occupied by dust bands of even smaller particles. The rings might be the result of a moon that has been blasted apart, and are thought to be no more 600 million years old.

THE MOONS OF URANUS

Uranus has 27 moons, five of which were photographed by Voyager 2 in 1986 although none has been comprehensively mapped. All orbit around Uranus' equatorial plane. This means that they, like the planet, are effectively lying sideways so their poles experience 42 years of night followed by 42 years of day. This made it impossible for Voyager to photograph the northern hemispheres of the moons, which were in darkness for the duration of its flyby.

All the major moons are composed of about half water-ice and half rock. Carbon compounds are thought to produce the moons' dark colours. All the major moons are tidally locked, so only one side would ever be visible from Uranus. The five major moons are all prograde (they rotate in the same direction as Uranus) and probably formed at around the same time as the planet. The minor moons are small – down to a few kilometres across – and many are retrograde and irregular. They might be captured asteroids, added to Uranus' entourage at later stages in the planet's history.

The five major moons of Uranus, and the largest of the minor moons, Puck. From top left, Titania, Oberon, Miranda, Puck, Umbriel, Ariel.

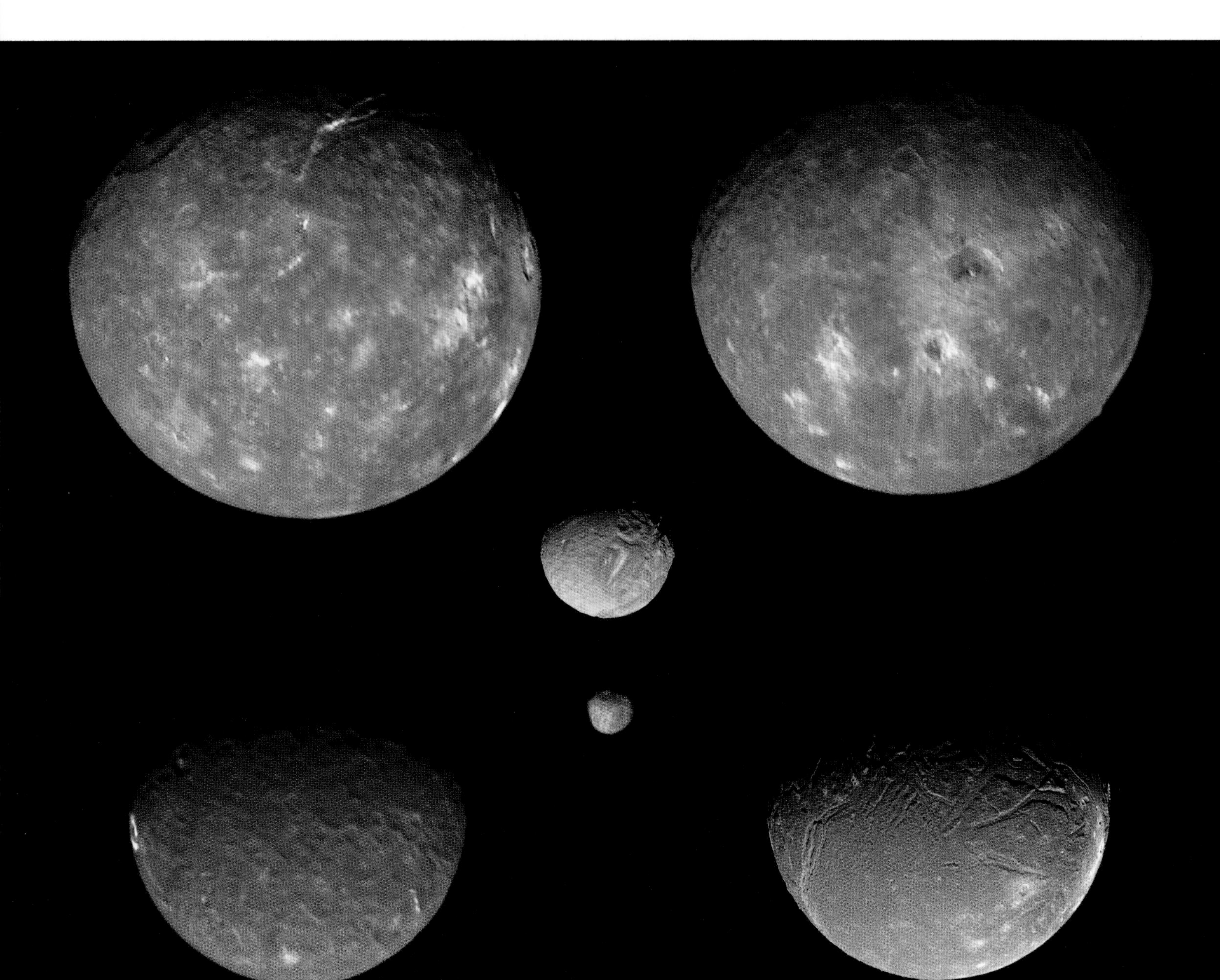

The sequence of Uranus' fourteen innermost moons, and their relation to two of the planet's rings. The moons are named after characters in the plays of William Shakespeare and Alexander Pope's poem The Rape of the Lock.

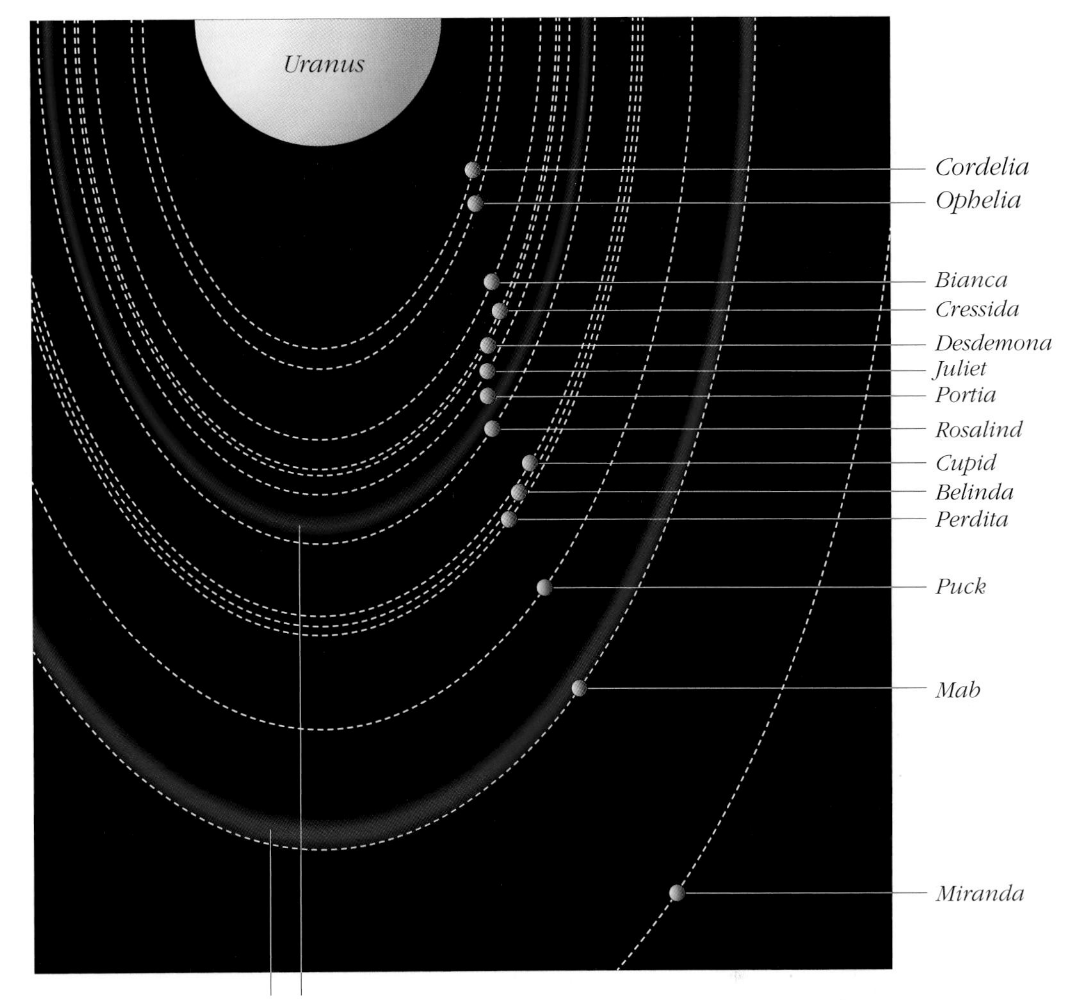

VOYAGER 2, ARIEL, 1986

This photograph of Ariel was taken by Voyager 2 from a distance of 130,000 km (80,000 miles). The surface, which is thought to be porous, is scarred by numerous craters caused by meteoroid impacts, but it is the youngest surface among the moons. The linear grooves are the result of tectonic activity. Only 35 per cent of the surface of Ariel was imaged by Voyager. It is the lightest-coloured, most reflective of the five large moons.

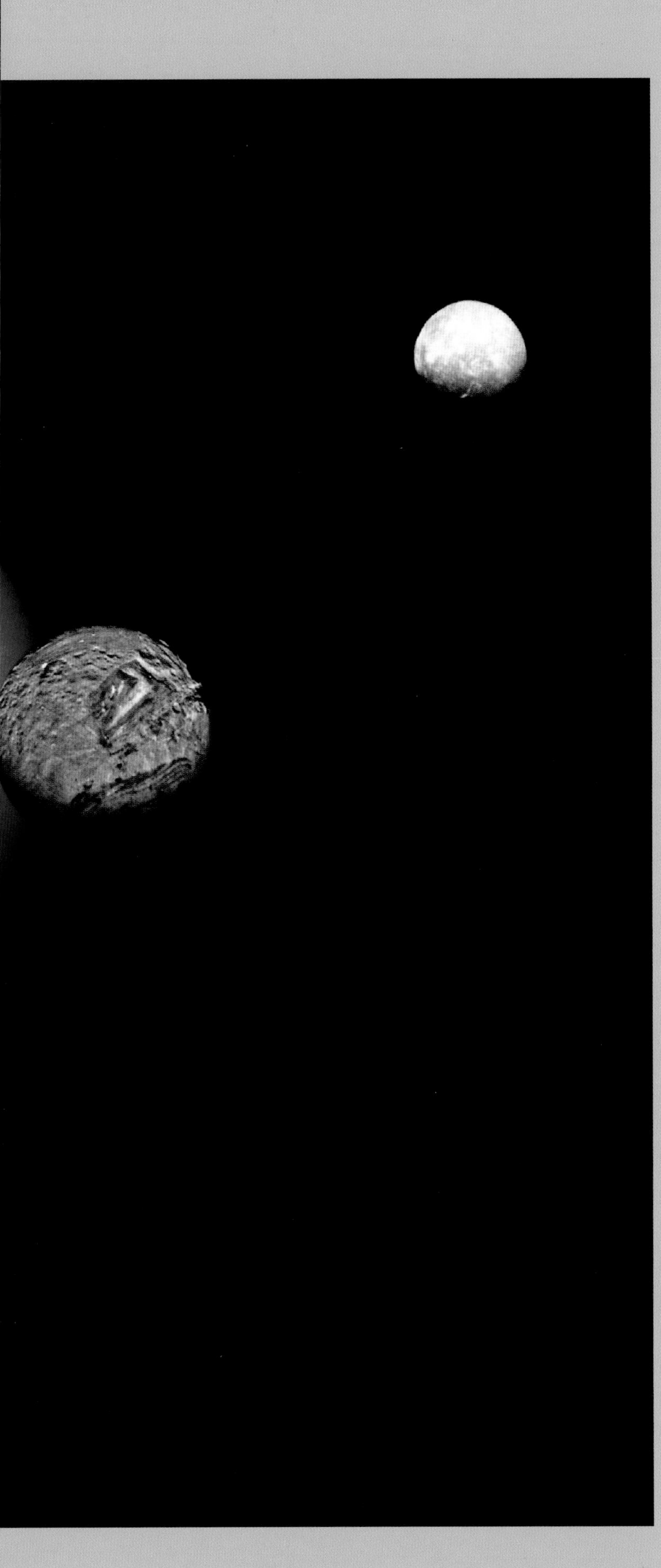

Left: *Uranus (centre) with five of its 27 moons.*

VOYAGER 2 & USGS, UMBRIEL, 1986

Umbriel is the darkest of Uranus' moons. Voyager 2 photographed 40 per cent of its surface, but only 20 per cent at sufficiently high quality for geological mapping. This USGS map of Umbriel (below) shows the cratered surface and its pattern of canyons and polygons. The polygons are oddly shaped areas of up to hundreds of kilometres across. They are found all over the mapped surface of Umbriel and thought to be evidence of some kind of internal geological activity long ago.

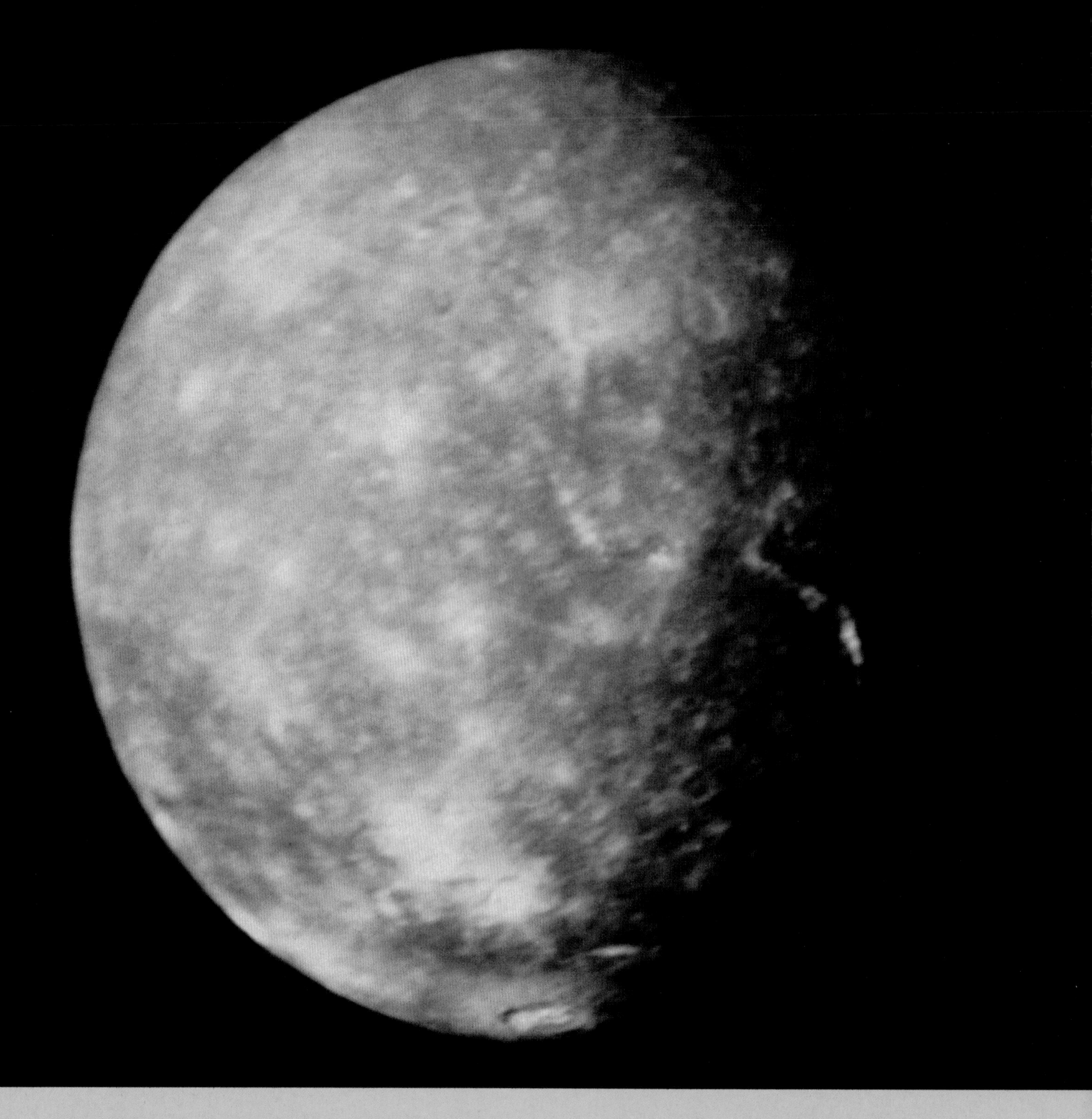

VOYAGER 2, TITANIA, 1986

Titania is the largest of Uranus' moons, at 1,600 km (1,000 miles) across. The trenchlike feature visible near the terminator (the line between light and dark) is caused by tectonic activity. There are many such scars on Titania's surface, the largest running from the equator almost to the south pole. They are typically 20–50 km (12–31 miles) across and up to 5 km (3.1 miles) deep and, as they run through craters, we know they are recently created features. The largest known crater, Gertrude, is 326 km (203 miles) across.

The surface of Titania is reddish, though less so than Oberon, Uranus' second-largest moon. The pale areas are reflective and probably represent water ice in valleys.

VOYAGER 2, OBERON, 1986

The second-largest moon of Uranus, Oberon, has at least one mountain that rises 6 km (3.7 miles) from the surface. As the radius of the planet is just 760 km (472 miles), that's equivalent to a mountain 50 km (31 miles) high on Earth – more than five times the height of Mount Everest. The reddest of the large moons, Oberon is thought to be overlaid with dust that has come from the irregular outer moons. The dust would be captured by the larger moon as it drifted slowly towards Uranus.

VOYAGER 2, MIRANDA, 1986

Miranda is the smallest of the major moons and the closest moon to Uranus. It has one of the most bizarre lunar landscapes in the solar system, with a strange collection of surface features and three large *coronae*, which are lightly cratered collections of ridges and valleys. These are separated from the more heavily cratered areas by sharp boundaries, giving the moon the appearance of something that has been pieced together from mismatched parts. It's not clear how this has come about. One suggestion is that the moon was in a collision which blasted it apart but the pieces stuck back together in a haphazard arrangement. Or they could be the result of massive asteroid strikes and the resultant melting of underground ice which then rose to the surface and refroze.

The canyons on Miranda are up to 12 times as deep as the Grand Canyon in North America. Some are 20 km (12 miles) deep – a substantial rift in a body only 470 km (290 miles) across. The 'chevron' is a prominent feature comprising a series of light and dark grooves, making a V-shape in the surface.

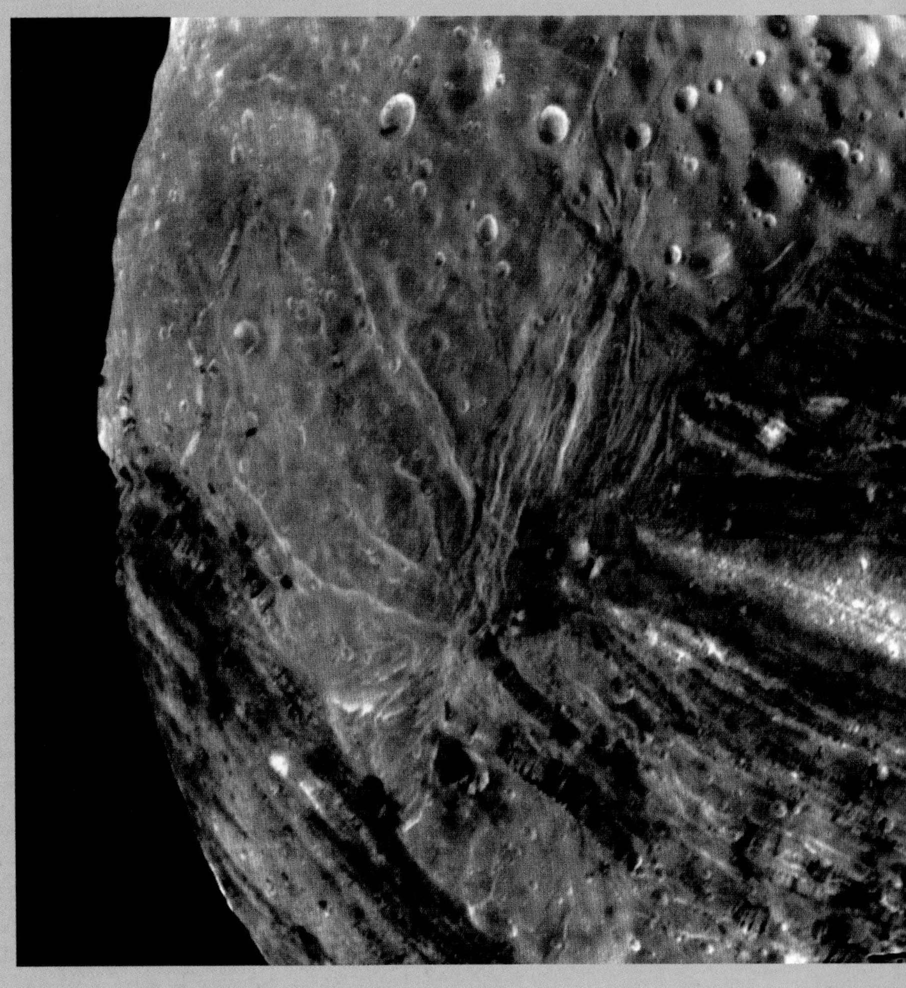

NEPTUNE

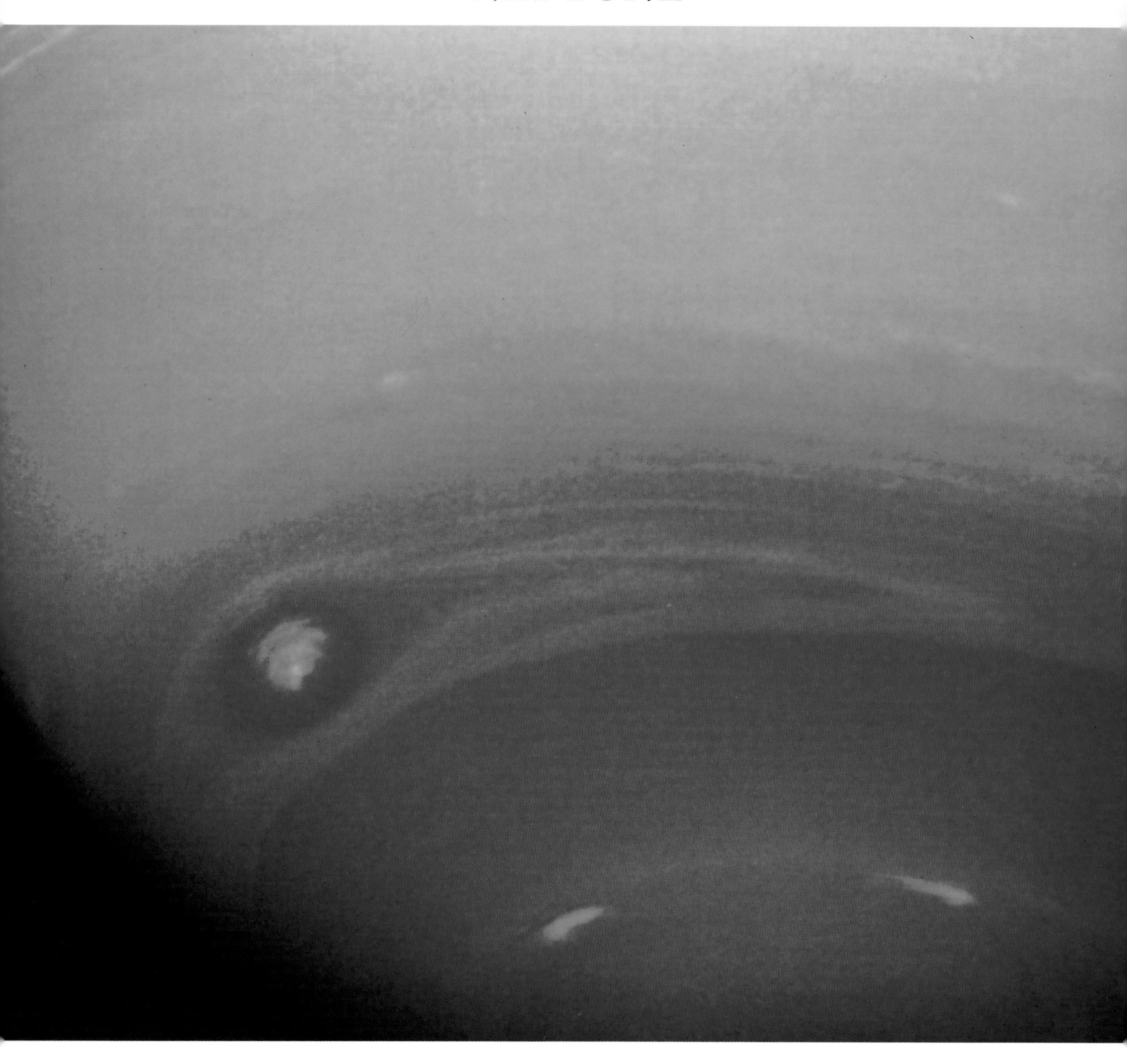

Neptune's beautiful blue surface with swirls of cloud looks serene, but the planet has the wildest winds of anywhere in the solar system.

Planetary focus	Neptune
Length of year	165 years
Length of day	16 hours
Size x Earth mass	17
Size x Earth radius	4
Average distance from Sun	4.5 billion km (2.8 billion miles)
Moons	14
Discovered	1846
Missions	Flyby: Voyager 2 (1989)

Neptune, the farthest planet from the Sun, has been observed up close only once, by the Voyager 2 spacecraft in 1989. It is smaller than Uranus, but has more mass. Like Uranus, it has a rocky core surrounded by a mix of water, ammonia and methane ices and an atmosphere of hydrogen, helium and methane. It looks even bluer than Uranus, which is probably the result of some other component not yet identified. Its clouds and storms are more readily visible – Voyager discovered a massive storm that appeared as a dark spot on the planet.

Neptune photographed at four-hourly intervals by Voyager 2 in 1989 shows the whole 'surface' over the course of a Neptunian day.

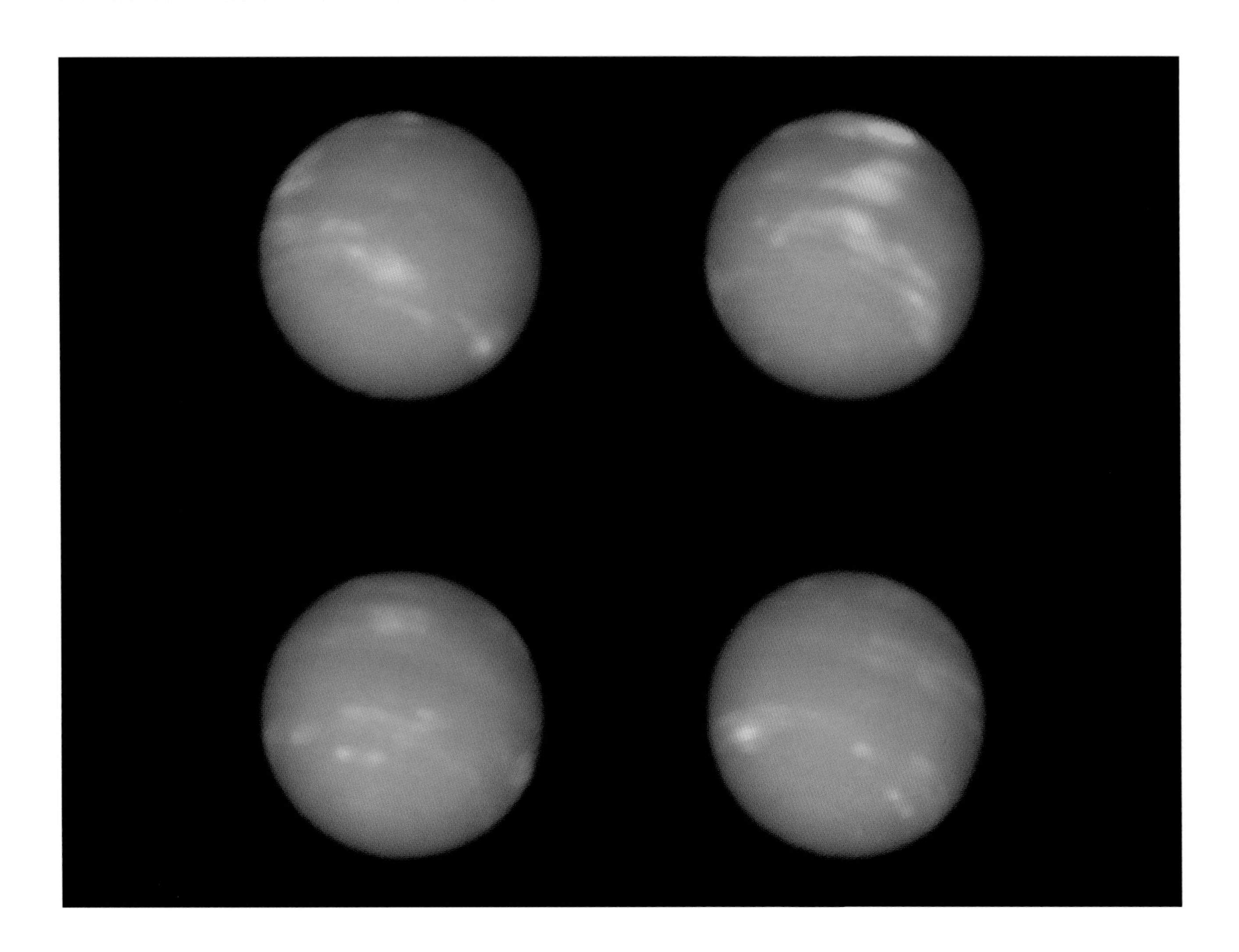

FOUND WITH MATHS

Neptune is too distant to be visible to the naked eye, and was only identified as a planet in 1846. Its distance from the Sun, and therefore its very slow progress across the sky, caused it to be mistaken for a star. It simply didn't move as the other planets do. Galileo saw Neptune as early as 1612. His drawings from observations on 28 December 1612 and 27 January 1613 show a 'star' in the position which would have been occupied by Neptune on that date; he even notes with surprise that two 'stars' (one of them Neptune) appeared to be further apart on the second occasion. Yet he didn't make the connection between the apparent movement and Neptune's status as a planet. Later astronomers also mistook Neptune for a star.

Neptune was finally recognized after mismatches between the observed and predicted orbit of Uranus led to the conclusion that the gravity of an unobserved planet must be affecting it. John Couch Adams in England and Urbain le Verrier in France both came to this conclusion and calculated the planet's orbit. Armed with this information, Johann Galle and his assistant Heinrich d'Arrest discovered Neptune on 23 September 1846, while working at the Berlin Observatory.

This diagram, published in The Illustrated London News *in October 1846, records the position of the newly discovered Neptune on two dates in September. It is marked in the first two complete squares in the upper left corner.*

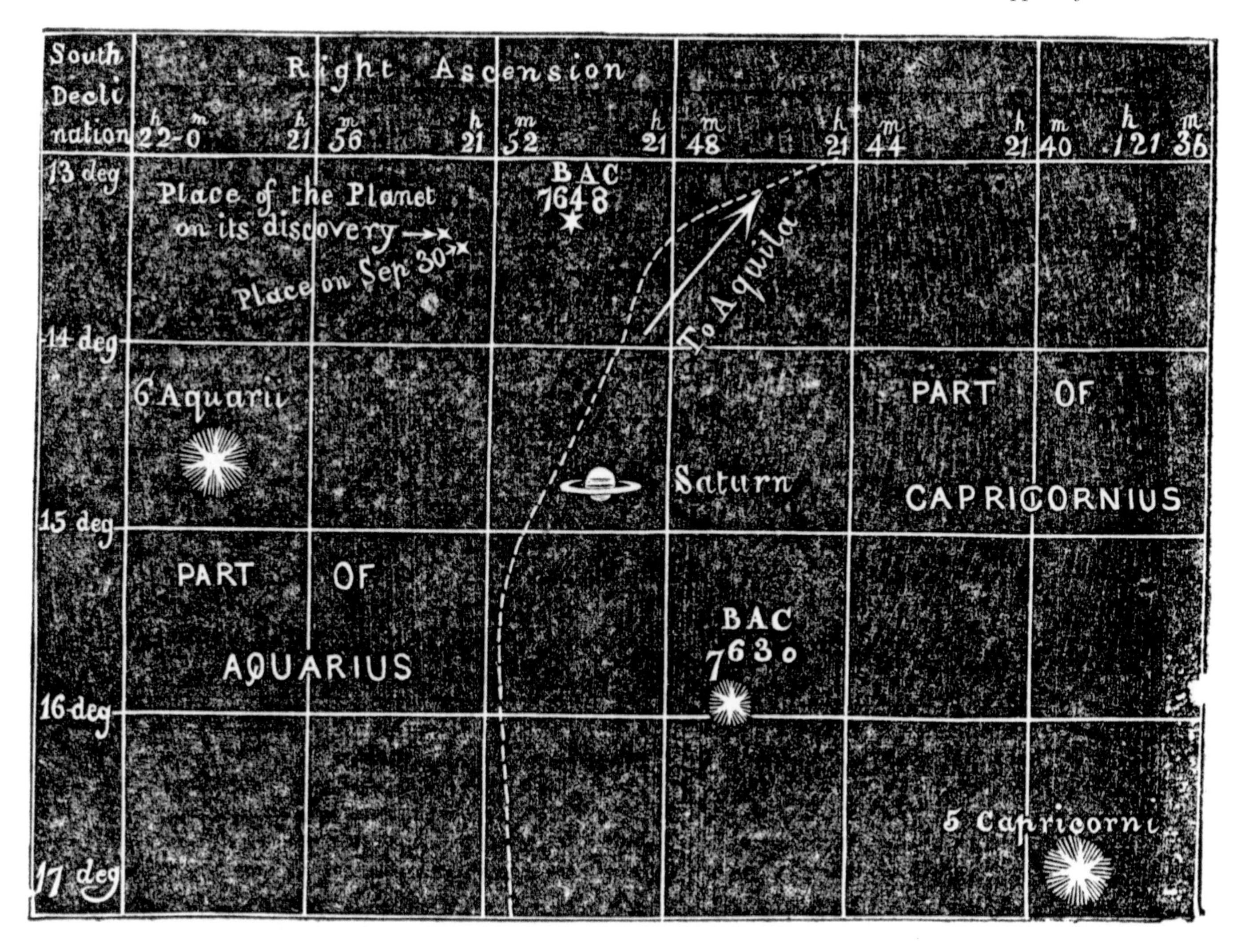

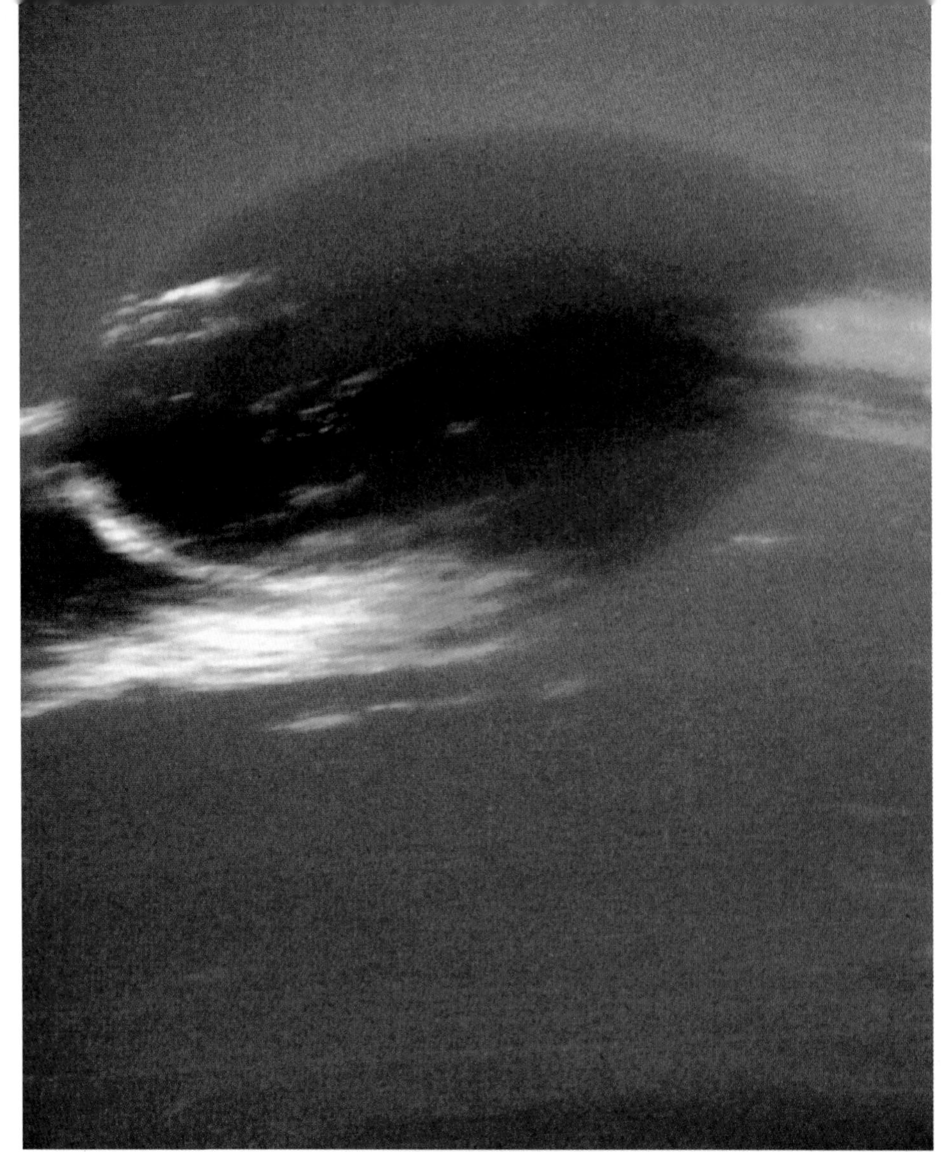

VOYAGER 2, NEPTUNE'S DARK SPOT, 1989

Voyager 2, the only spacecraft to approach Neptune, recorded a vast storm over the planet's southern hemisphere, which was soon named the Dark Spot. It had the strongest winds ever recorded on any planet, at up to 2,414 km per hour (1,500 mph). The Spot persisted for about five years, but had gone by the time the Hubble Space Telescope observed the planet in 1994. Hubble did reveal a smaller storm in the northern hemisphere of Neptune. Even during the brief period of Voyager's observations, the Spot changed.

VOYAGER 2, CLOUDS ABOVE NEPTUNE, 1989

Winds whip around the planet at up to 600 m (1,970 ft) per second carrying clouds and producing storms. Hydrocarbon haze in the upper atmosphere probably freezes into snow that melts as it falls into lower levels, heated by increasing atmospheric pressure. This photograph taken by Voyager makes it clear that the white cloud lies on top of the atmosphere.

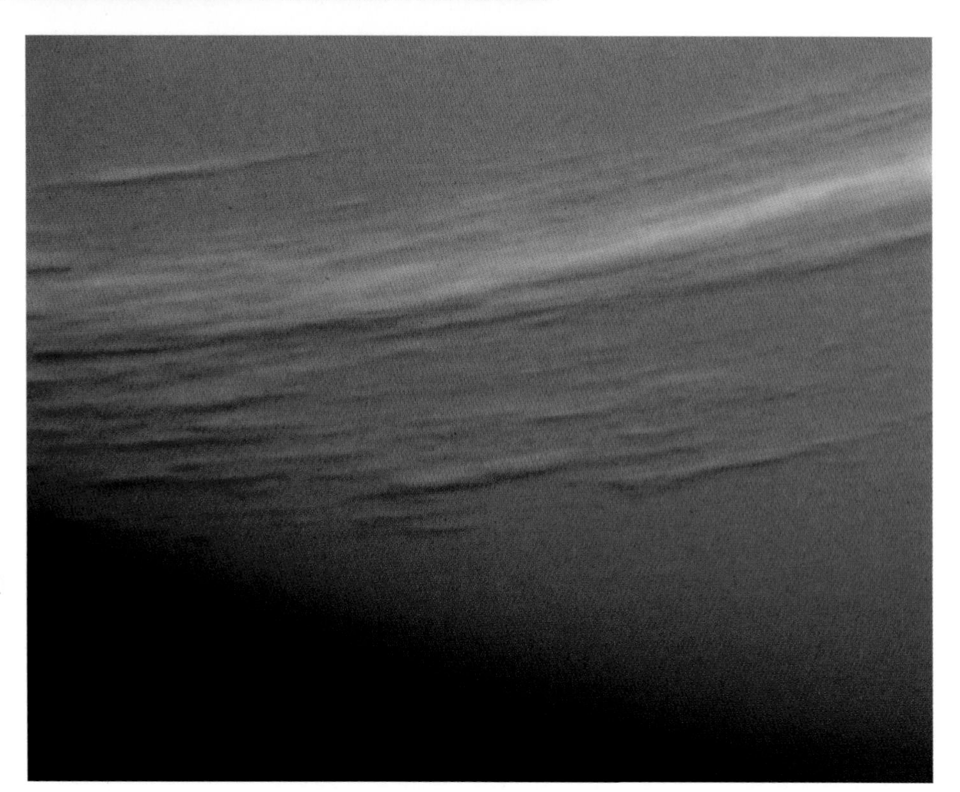

VOYAGER 2, NEPTUNE'S RINGS, 1989

Neptune has a faint ring system, suspected on the basis of observations by telescope but fully confirmed by Voyager 2 in 1989. There are two narrow, bright rings: the innermost, Le Verrier, is 113 km (70 miles) across and the outer Adams ring is an average of 35 km (22 miles) across. Both are only a few hundred metres thick. Two faint, much broader rings are 2,000 km (1,243 miles) and 4,000 km (2,485 miles) across and, respectively, 150 m and 400 m thick. The outer edge of the widest ring is brighter and is sometimes named as a separate ring, 100 km (62 miles) wide. The ring material is very dark, and has a high proportion of dust (very small particles). It is probably a mix of water ice, rock and some carbon-based compounds.

The two brightest rings clearly visible in this image are the Le Verrier and Adams rings.

NEPTUNE'S MOONS

Neptune has 14 moons, five of which were discovered by Voyager in 1989. The largest, Triton, was spotted just 17 days after Neptune was identified in 1846. The second to be found was one of the planet's outermost moons, Nereid, in 1949. No more were detected until the 1980s. Hippocamp was recently discovered (in 2013) from images taken in 2004. The moons range in size from about 2,700 km (1,678 miles) to just 35 km (22 miles) in diameter.

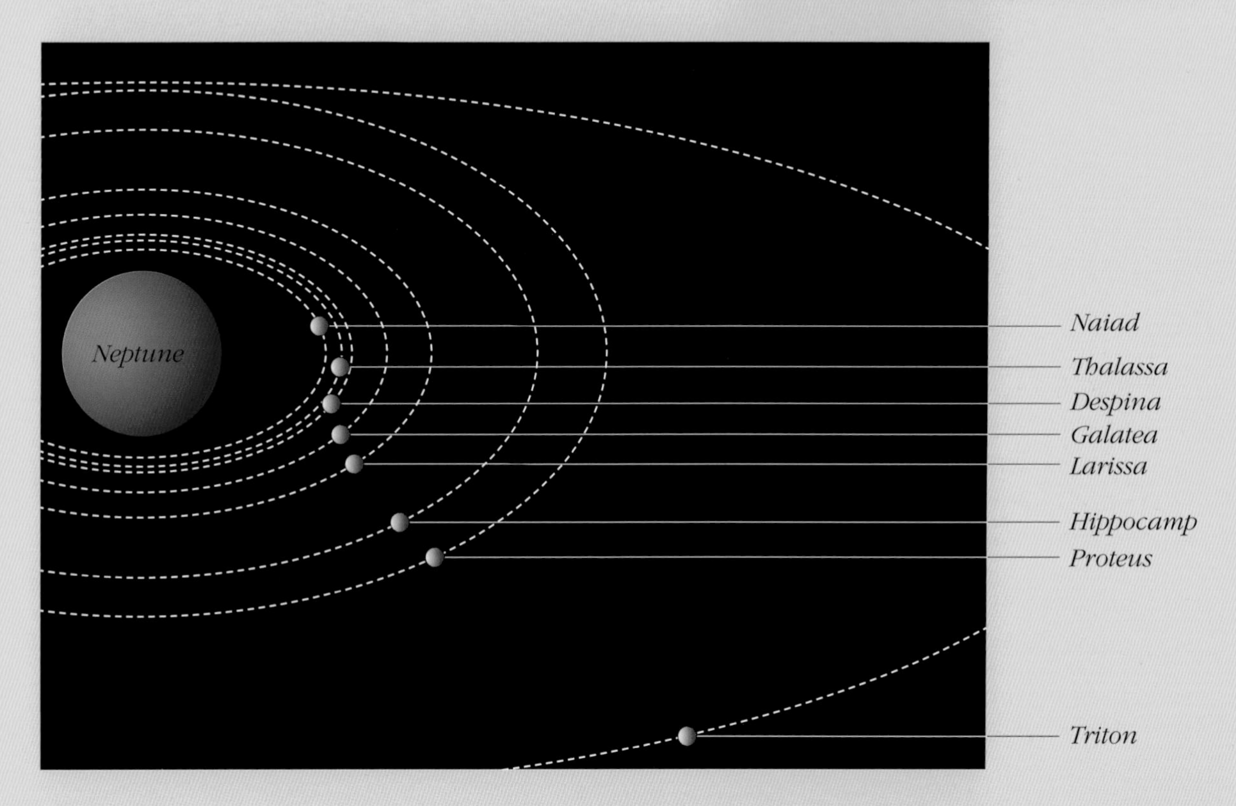

Moons of Neptune larger than 150 km in diameter at 1 km/pixel
Data from Voyager 2 courtesy NASA/JPL. Processed images and collage Copyright Ted Stryk

VOYAGER 2, THE LARGEST MOONS, 1989

Many of Neptune's moons have been recycled. The original clutch of moons was catastrophically disrupted at some point in the planet's history when it captured its largest moon, Triton. The first moons probably collided, forming a massive rubble pile from which new moons accreted. The two outermost moons, Psamathe and Neso, orbit further from their planet than any other known moons in the solar system.

Seven inner moons are regular and follow prograde circular orbits; seven outer moons are irregular and some follow retrograde orbits. Voyager 2 captured high quality images only of Triton (left), shown here alongside Neptune's other largest known moons (right).

Voyager 2's photos of Triton reveal a lightly cratered surface with extensive smooth plains. Triton is one of only four geologically active bodies in the solar system; the other three are Earth, Venus and Io. Its surface is covered with a frost of frozen nitrogen over an ice sheet, making the moon brightly reflective; beneath is a core of rock and metal. A thin atmosphere of nitrogen and methane is produced by geysers, but with a surface temperature of just -235 °C (-391 °F), this readily freezes.

TRITON, DESTROYER OF MOONS

Triton is one of the most mysterious moons in the solar system. Unlike other major moons, its orbit is retrograde. It's orbit is also highly inclined (tilted with respect to the plane of Neptune's equator), almost circular, and very close to the planet. Triton alone accounts for 99.7 per cent of the mass of Neptune's entire moon system. These peculiarities can be explained if Triton was originally a Kuiper Belt object captured by Neptune.

Triton is implicated in the demise of other moons around Neptune. Modelling suggests that if Neptune originally had a more extensive collection of moons comparable to that of Uranus, the capture of Triton would have destroyed many of them, resulting in a system that resembles that which Neptune has today. The present moons might be a mix of survivors and new moons formed from the debris of those blasted apart. Some, such as Naiad and Nereid, are thought to be piles of rubble rather than single, solid bodies.

NASA, HIPPOCAMP, 2019

The artist's impression of Hippocamp (below) was created after the moon's likely origins were worked out in 2019. It is thought to be a chunk knocked off the nearby moon Proteus in a collision with a comet four billion years ago. Hippocamp was never in the right place to be spotted by Voyager 2, and is very faint: only a tiny speck in the best telescopic photographs. It has just one 100-millionth the brightness of the dimmest star visible to the naked eye. Its discovery suggests there may be more unidentified moons in the solar system that are chips off a larger moon. Hippocamp orbits Neptune in just 22 hours.

BEYOND THE PLANETS

ON THE EDGES OF THE SOLAR SYSTEM

There are more bodies in the solar system than planets and moons, including a host of 'dwarf planets', or planetoids. There are also smaller asteroids, Kuiper Belt objects and comets. In a barely explored region beyond the orbit of Neptune, the Sun is circled by a disk populated by dwarf planets and rocky and icy objects. The disk is up to 20 AU wide – twenty times the distance from Earth to the Sun.

Left: *The rocky planets of the inner solar system are surrounded by the band of rubble that is the Asteroid Belt, lying between the orbits of Mars and Jupiter (top). But this is just the small centre of the entire solar system, which extends beyond the gas and ice giants to a vast ring of icy bodies and dwarf planets called the Kuiper Belt (below).*

WORKING WITH THE LEFTOVERS

The distinction between planets and not-planets is one made by humans – the solar system doesn't recognize any such distinction and there are no natural boundaries between categories of objects.

As the solar system formed, gravity acted even on the matter that was not drawn into planets and circumplanetary disks. This, too, formed clumps. Some of these clumps grew nearly big enough to qualify as planets. They accumulated sufficient mass to become roughly spherical, but not enough to clear their orbital paths of other smaller objects. The most famous of these dwarf planets is Pluto, which until 2006 was classified as a planet. There are four other confirmed dwarf planets: Ceres, Haumea, Makemake and Eris. Of these, Ceres alone is in the asteroid belt, between Mars and Jupiter. The others orbit beyond Neptune, so are also classed as trans-Neptunian objects (TNOs). In addition there are candidate dwarf planets such as Sedna, Quaoar and Biden. Some dwarf planets are believed to be the cores of failed planets.

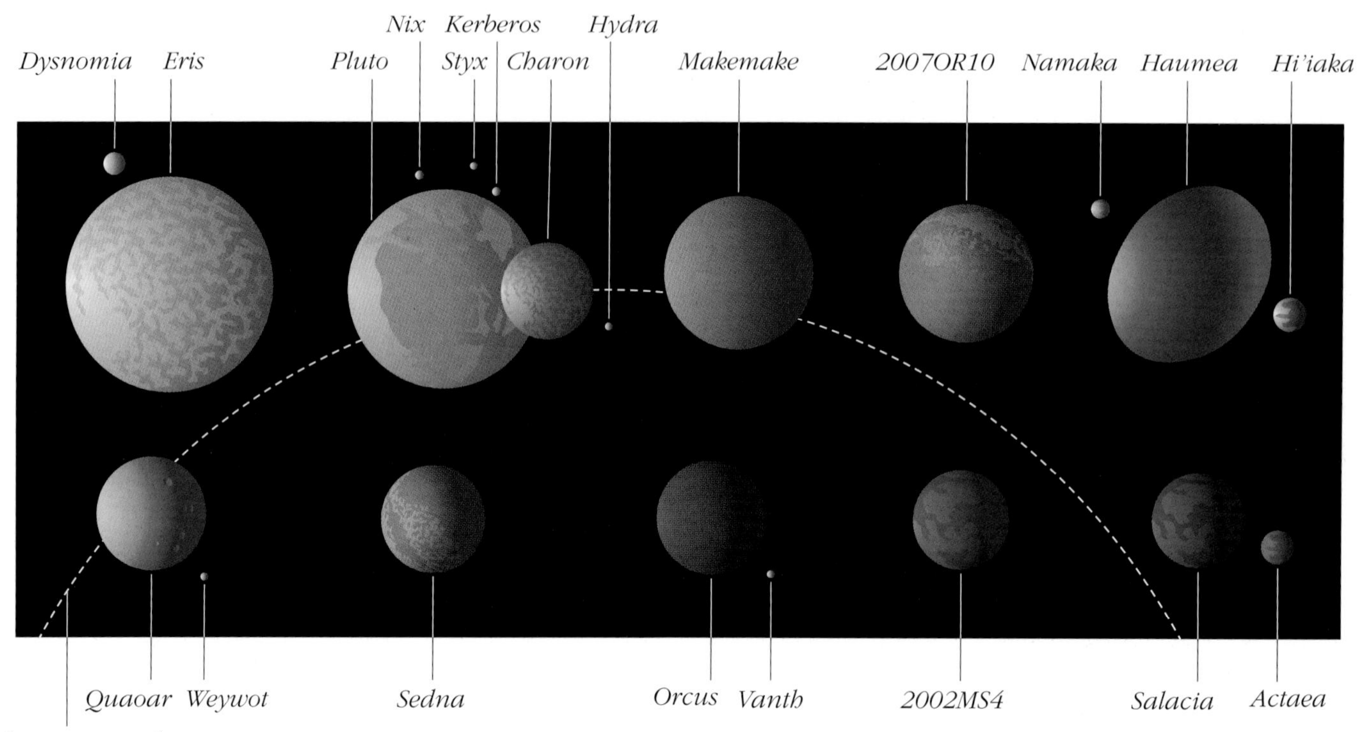

Large objects in the Kuiper Belt and their moons. The largest known objects – Pluto, Eris, Makemake and Haumea – are dwarf planets. 2007OR10 is the third or fourth largest object and could be classified as a dwarf planet. It is currently the largest body in the solar system without a name.

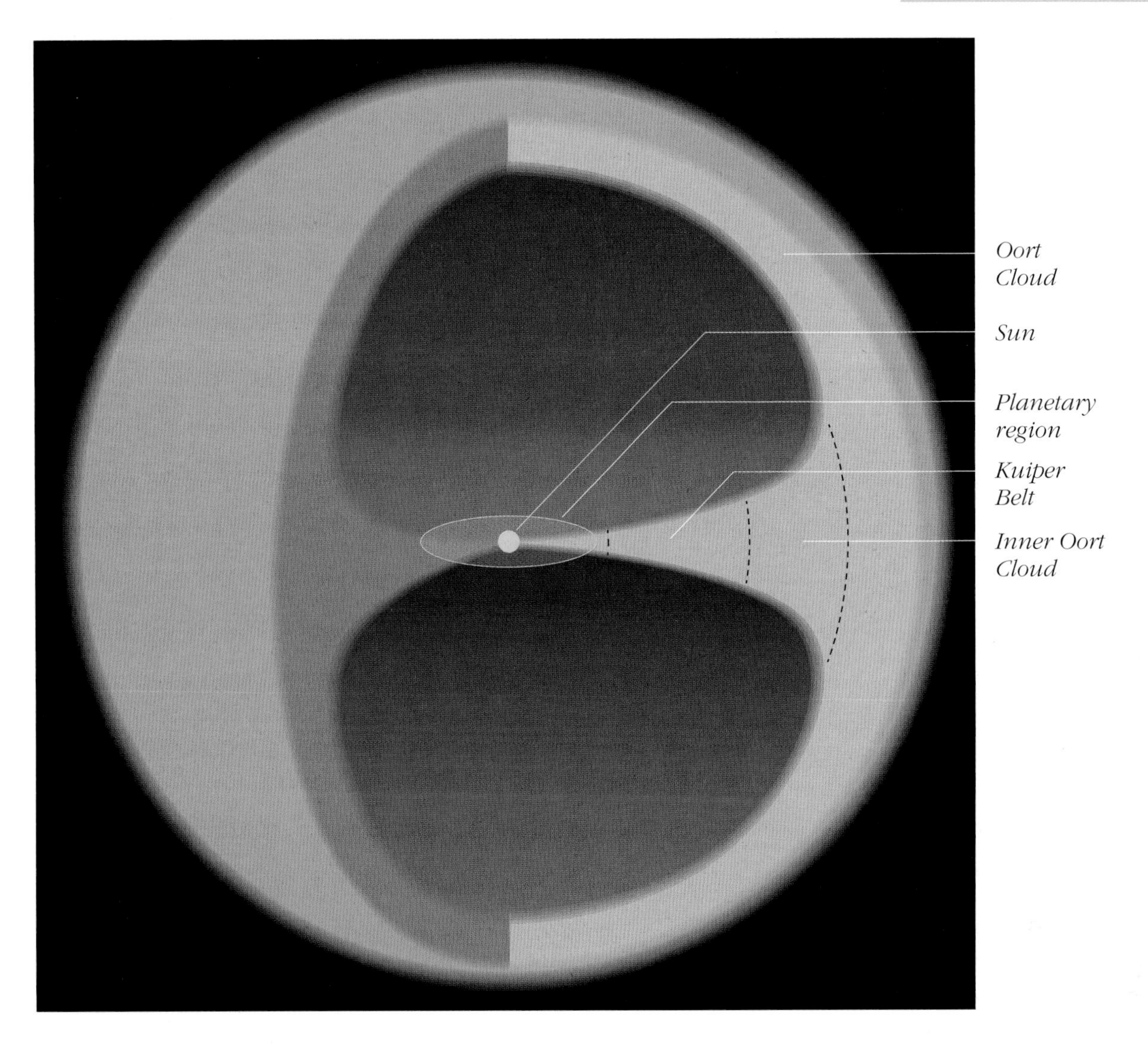

Above: *The Kuiper Belt is a region beyond the orbit of Neptune populated by billions of icy bodies of varying sizes. Beyond that, the Oort Cloud, a spherical shell of more icy objects, surrounds the entire solar system. It begins about 1,000 AU from the Sun and extends more than 100,000 AU. It's thought to contain at least two trillion objects.*

Below: *The orbit of Pluto (yellow, inclined circle) in the Kuiper Belt, a doughnut-shaped region of icy bodies beyond the orbit of Neptune.*

COMETS FROM THE KUIPER BELT AND BEYOND

The Kuiper Belt is the source of many comets – icy bodies on long elliptical orbits which become visible as they approach the Sun. Comets with a return period of less than 200 years originate in the Kuiper Belt. Comets with a longer return period – up to thousands or even millions of years – are thought to come from the Oort Cloud.

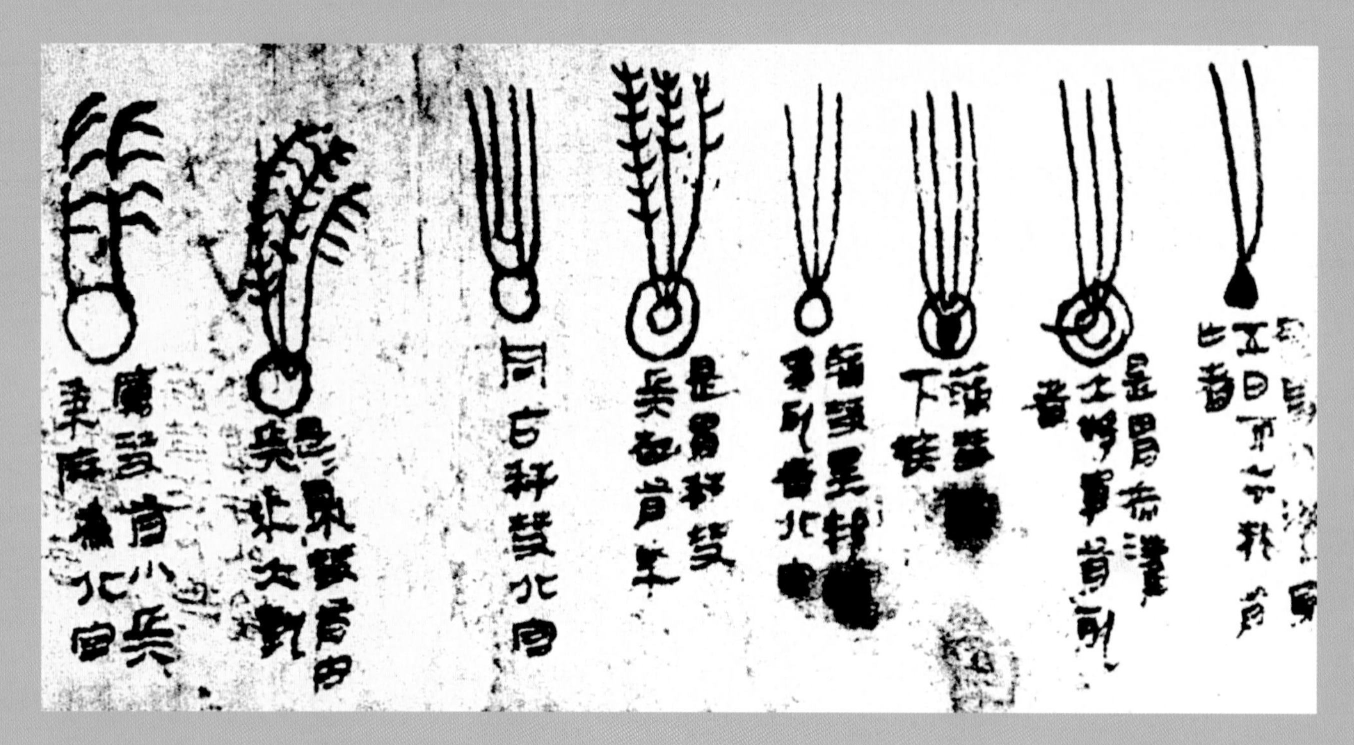

Above: *A silk manuscript found in a tomb from the 2nd century* BC *shows seven comets observed by Chinese astronomers. This is the earliest surviving depiction of comets in any detail.*

Right: *A section of the Bayeux Tapestry, made in Normandy, France in the 1070s, shows people pointing at a comet that was believed to be an omen. It appeared in 1066 and is thought to be Halley's comet.*

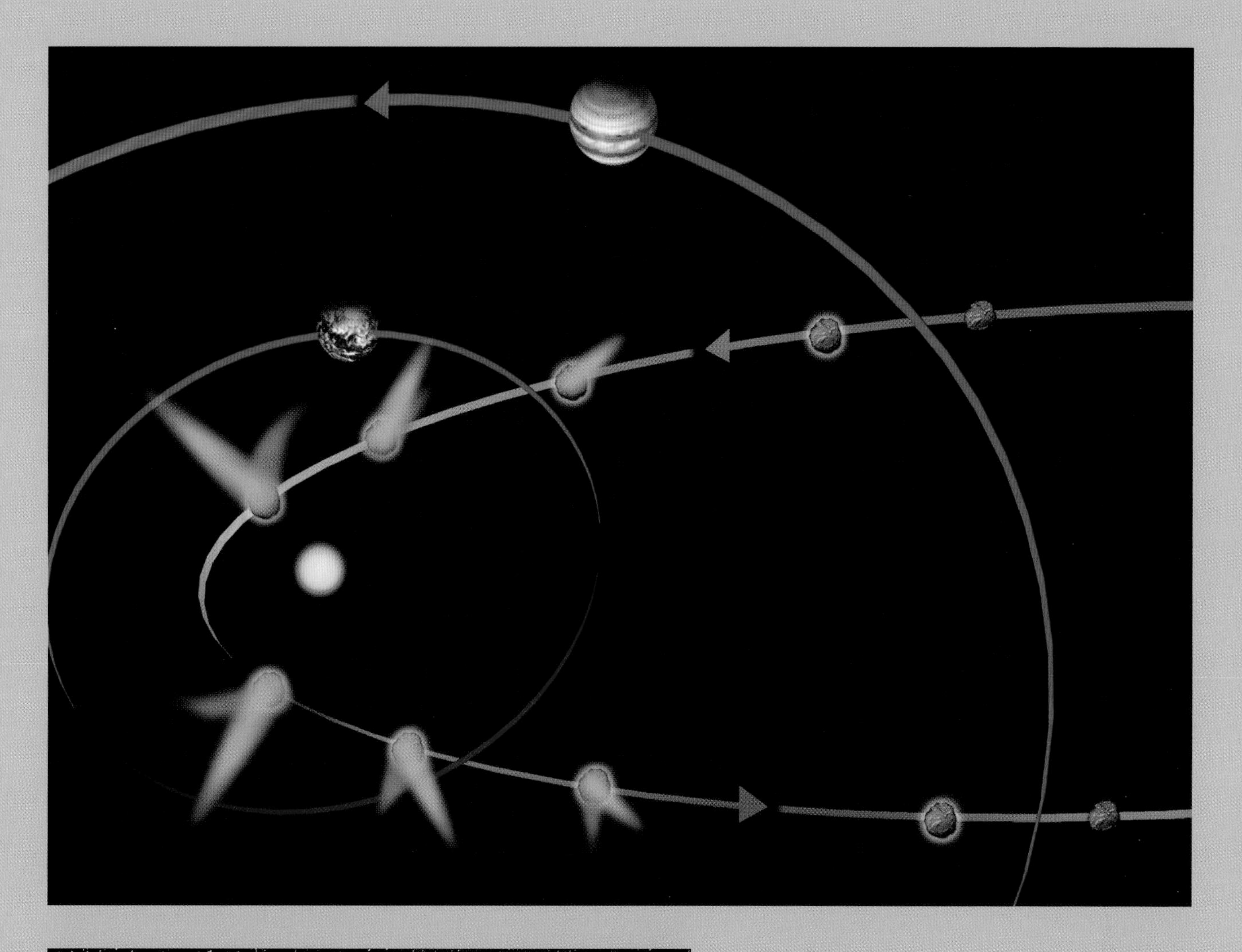

Above: *As a comet's nucleus approaches the Sun (following the green trajectory from the right), it begins to heat up. The first effect is that it develops a coma, an envelope of gas. Approaching Earth's orbit, it develops a tail of ionized gas, produced by solar radiation ionizing gases in the coma. Even closer to the Sun, the comet develops a tail of dust – rocky particles freed from the nucleus as the ice binding it melts and evaporates. Then, as it moves away, it loses first its tails and then its coma, returning to a cold, icy lump, just slightly diminished by its tour around the Sun.*

Left: *The comet Hale-Bopp, photographed in 1997. The ion (blue) and dust tails are clearly visible.*

THE DWARF PLANETS

The term dwarf planet was introduced in 2006 when the discovery of more objects about the size of Pluto threatened to overburden the solar system with planets. All planets, including dwarfs, become roughly spherical under the influence of their own gravity as they spin. All planets and dwarf planets must be in orbit around the Sun and not around anything else (that is, they must not be moons). But while major planets clear debris out of their orbit – they either absorb it or it is clumped into moons – dwarf planets are too small and don't have sufficient gravity to accomplish this. Although only five dwarf planets are currently recognized officially, it's likely there are around 200 of them in the Kuiper Belt.

Right: *Ceres, the first dwarf planet discovered, is the only one to be located in the Asteroid Belt.*

Unlike asteroids, dwarf planets don't have the same composition all the way through. The heavy materials (rock and metals) have sunk to the core while lighter materials (typically ice and organic compounds) are nearer the surface. The first dwarf planet discovered was Ceres in 1801 – even before Neptune was recognized. It was classed as a planet for around 50 years, then reclassified as an asteroid until 2006. The existence of Pluto was predicted in 1915, based on research into its effects on the orbits of Uranus and Neptune. Pluto itself was found in 1930 by Clyde Tombaugh. The remaining three dwarf planets were discovered this century – Haumea in 2004, and Makemake and Eris in 2005.

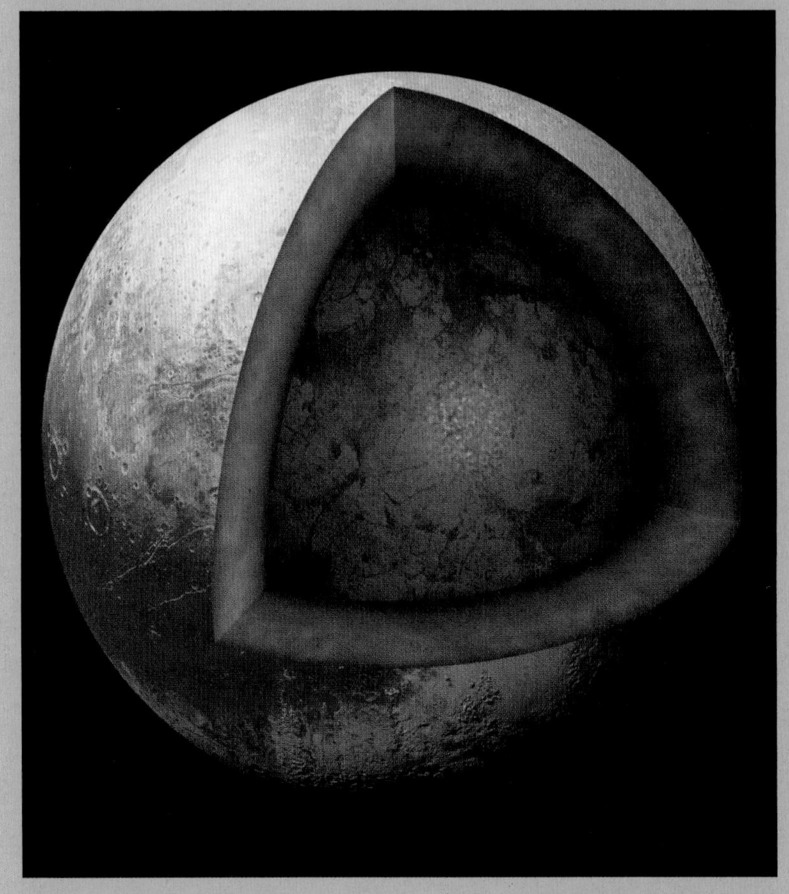

Left: *Data from the Hubble Telescope suggests that Pluto's rocky core occupies 70 per cent of its diameter. It's surrounded by a mantle of ice under a crust of water ice mixed with methane, nitrogen and carbon. There's possibly an ocean of liquid water and ammonia above the core.*

NASA, THE HEART OF PLUTO, 2015

Pluto's rotation is retrograde and its axis is tilted at around 120°, so its south pole is further 'up' than its north pole. It has a layer of frozen nitrogen and water over a thick mantle of water ice and a large rocky core. NASA's New Horizons spacecraft took this photograph from a distance of 768,000 km (477,000 miles). The most prominent feature is the 'heart' of Pluto's northern hemisphere, formally known as *Tombaugh Regio* after astronomer Clyde Tombaugh. The region is 1,600 km (1,000 miles) across. The dark colour of the surface beside the heart is thought to be the result of tholins – complex carbon compounds created by the action of sunlight on methane. The smooth appearance of *Tombaugh Regio* suggests that ongoing or recent geological activity might have covered any features previously present. The whole area is rich in nitrogen, carbon monoxide and methane ices. The heart was visible as a bright area through telescopes before New Horizons' visit, and has been dimming for around 60 years. Young mountains on the planet are further evidence of recent geological activity.

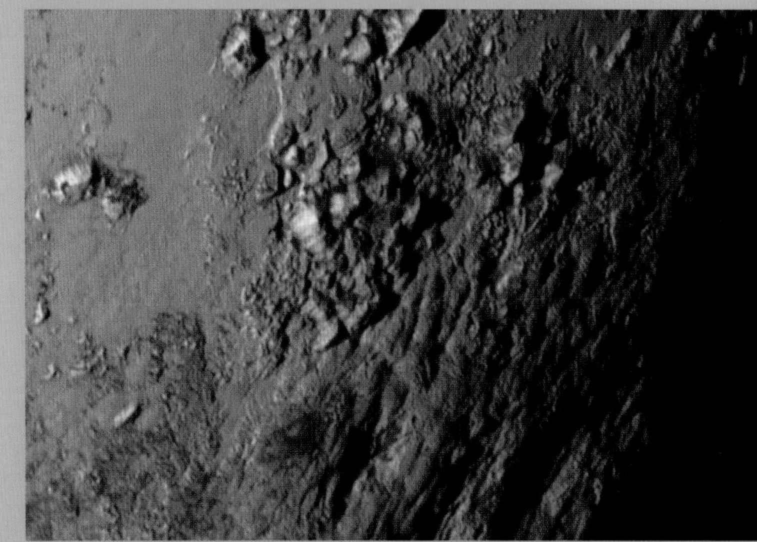

This range of mountains rises 3,500 m (11,000 ft) from the surface of Pluto. The mountains are recent – perhaps formed just 100 million years ago, and maybe still forming. The peaks are probably made of water ice, which forms the bedrock of Pluto's surface.

MOONS LARGE AND SMALL

Pluto has five moons, of which Charon is by far the largest. In relative terms, at almost half the diameter of its host, Charon is larger than any other moon in the solar system. All five of Pluto's moons probably formed as the result of a collision between Pluto and another Kuiper Belt object of similar size.

Charon was discovered in 1978; the other moons were all found with the Hubble telescope – Nix and Hydra in 2005, Kerberos in 2011 and Styx in 2012. The smaller moons are too small to be spheroid, and tumble chaotically in their orbits so there is no standard-length day on any of them.

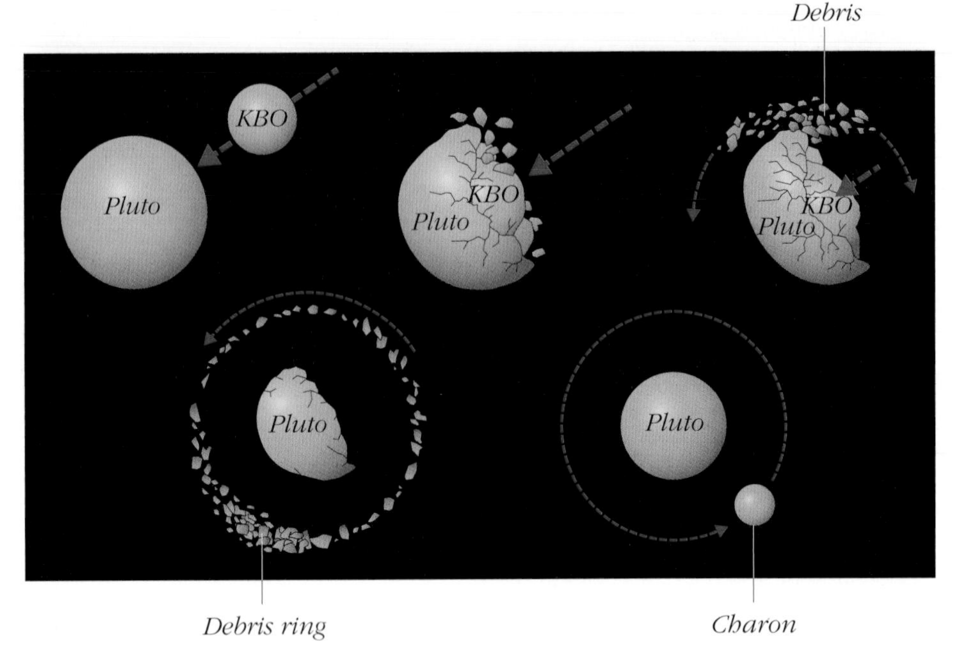

Above: *The process by which Pluto's largest moon, Charon, probably formed. An impactor destroyed a large part of Pluto, which became a ring of debris around the deformed planet before gravity re-rounded Pluto and formed a moon from the debris.*

Below: *We have only blurry photos of Pluto's four smaller moons – but it's astonishing that we can see this much since the smallest, Styx, is just 16 km (10 miles) across but 7.5 billion km (4.7 billion miles) away! Kerberos and Hydra might each be the result of two smaller moons combining.*

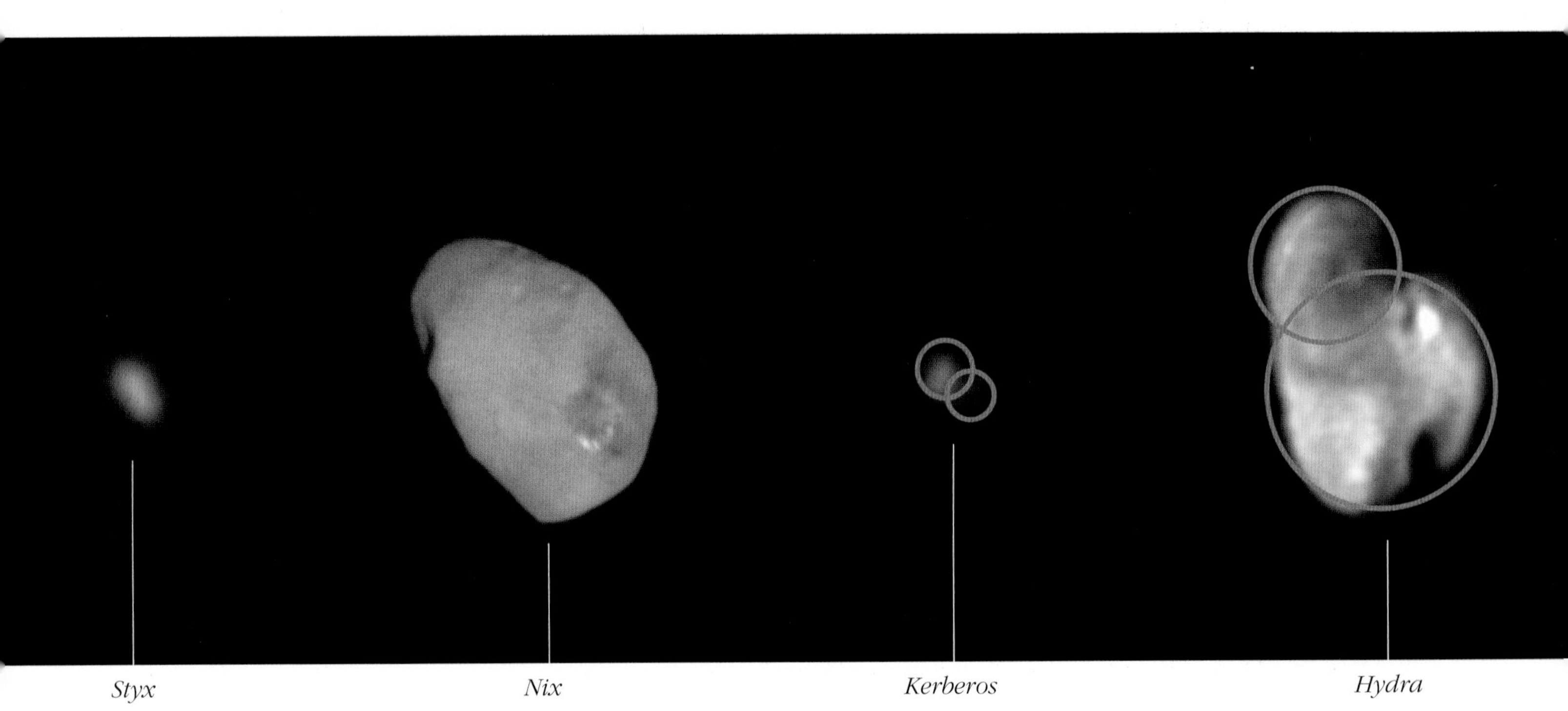

NEW HORIZONS, CHARON, 2015

Although technically a moon of Pluto, Charon is so large that some astronomers refer to Pluto–Charon as a two-body system. It lies very close to Pluto, at just 19,640 km (12,200 miles) – less than the distance by plane from London to Sydney. Its orbit around Pluto takes 6.4 days and it is tidally locked.

Charon's surface is mostly grey and covered with water ice, though an area around the north pole has a reddish tinge (exaggerated in this colour-enhanced image). The source of the colour is thought to be from shreds of Pluto's atmosphere. Methane falls from Pluto onto Charon, sticks to the cold surface of the pole at night and forms tholins under the action of ultraviolet. The tholins react with frozen water on the surface to produce the red compound.

NEW HORIZONS, WRIGHT MONS, 2015

The surface of Charon poses puzzles for astronomers. Its smoothness and relative lack of craters suggests surface renewal of the type produced by cryovolcanism (volcanic eruptions occurring at low temperatures), yet the moon is too small and cold for geological activity. It's possible that one or more collisions have melted the surface, resulting in smoothing. But the massive network of canyons and cliffs, extending up to 1,600 km (1,000 miles) and reaching depths of 7.5 km (5 miles), suggests instead a collision that has cracked but not melted a solid surface. Such depressions, or chasmata, form a band around the equator of Charon.

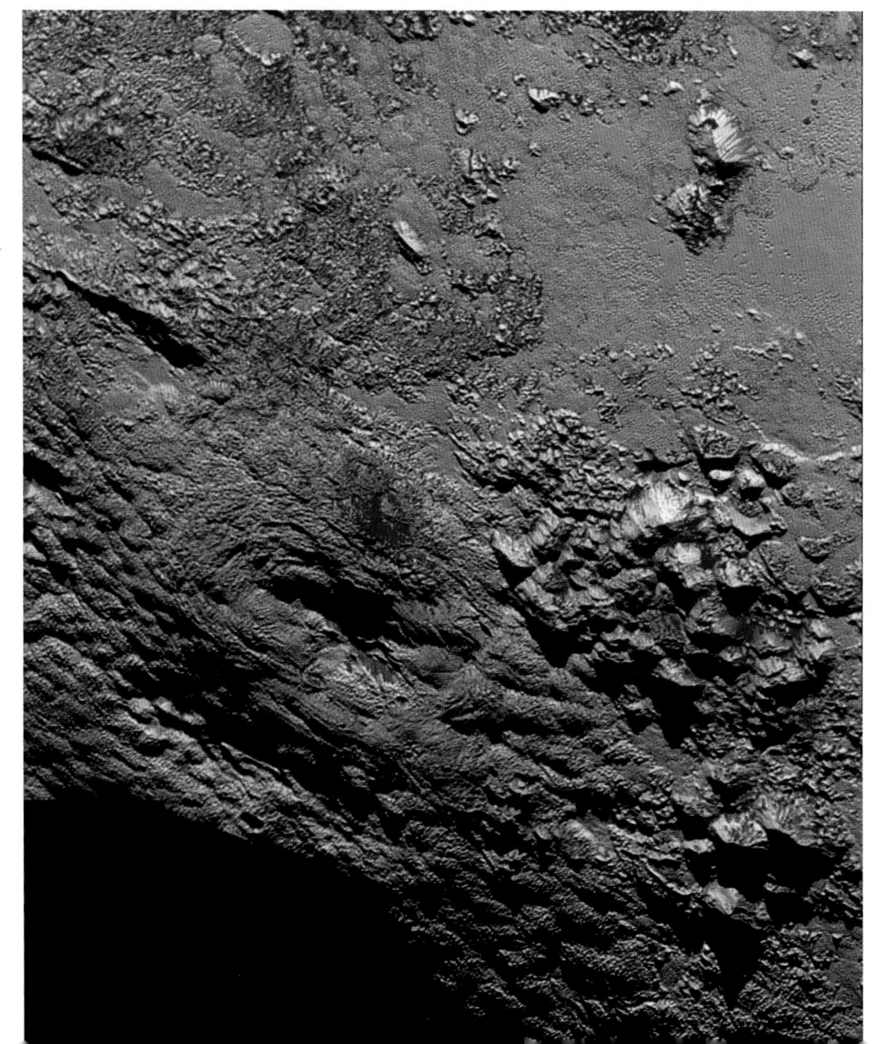

NASA & ESA, ERIS, 2006

A dwarf planet with a highly elliptical orbit that takes 558 years to complete, Eris is the largest known KBO apart from Pluto. Discovered in 2005, it has a bright, reflective surface, as shown in this artist's impression. Eris has one known moon, called Dysnomia.

Eris and its moon, Dysnomia, (below left); and an artist's impression of Eris and Dysnomia with the Sun in the far distance (below right).

NASA, MAKEMAKE, 2016

This artist's impression of Makemake focuses on the reddish brown colour of the dwarf planet. It probably has a layer of frozen methane on it surface, maybe made of small pellets 1 cm (0.5 in) across. There is also evidence of frozen ethane and nitrogen. It has a single moon, nicknamed MK2, about 160 km (100 miles) across. Its day is 22.5 hours long and its year 305 years long.

JOHN R. FOSTER, HAUMEA AND ITS MOONS

Little is known about the strangely egg-shaped dwarf planet Haumea (facing page). Discovered in 2005, it is the largest body in a group of objects thought to be the result of a huge impact billions of years ago. A large moon probably formed after the collision, but it has since been destroyed. Haumea has the highest rotation rate of any celestial body in the solar system wider than 100 km (62 miles). It completes its 'day' in just under four hours (though its year is 285 Earth years long, meaning that there are nearly 625,000 Haumea days in one Haumea year). It has two moons, Namaka and Hi'iaka, and a simple ring system of dust and ice crystals about 70 km (44 miles) wide, making it the first-known KBO to have rings.

THE KUIPER BELT AND ITS OBJECTS

The Kuiper Belt is a broad band (about 20 AU across) that begins 30 AU from the Sun. It is a debris field, littered with lumps left over from the formation of the solar system. It's a bit like the Asteroid Belt, but much bigger: twenty times as wide and with up to 200 times as much mass. It's thought there might be no clear distinction between small objects in the solar system, but that they fall on a spectrum with those containing more metal or rock inhabiting the Asteroid Belt and those containing more ice inhabiting the Kuiper Belt. Centaurs, which orbit between the Asteroid Belt and Neptune, have a composition between the two, mixing rock and ice.

The first Kuiper Belt Object, or KBO, was discovered in 1992; the second followed six months later. Now more than 1,000 are known and it's thought likely that there are 100,000 KBOs larger than 100 km (62 miles) across. There could be billions that are smaller. The smallest so far discovered is just 1.3 km (0.8 miles) across and was found by an amateur team in Japan using cheap telescopes.

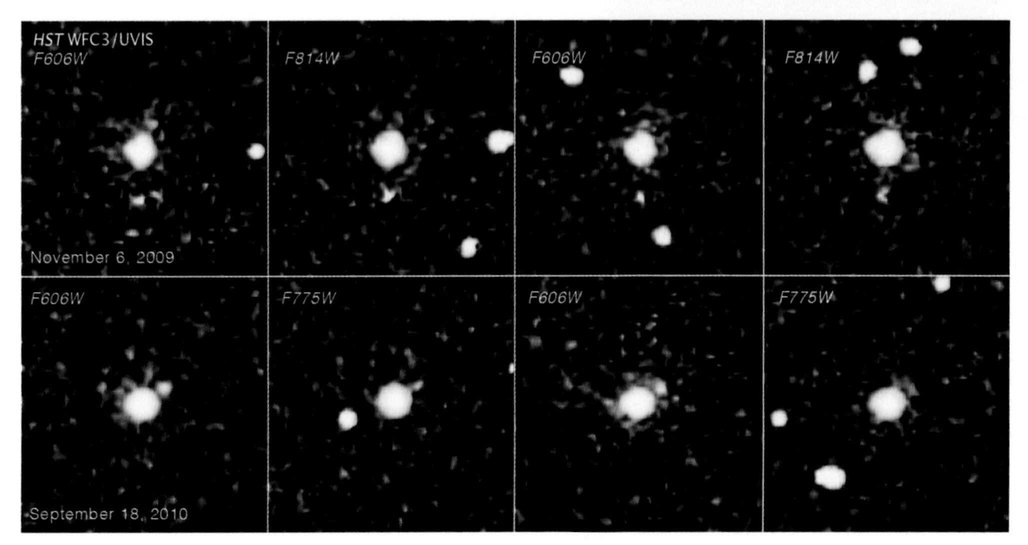

Spectroscopy suggests that KBOs are made of various ices, including methane, ammonia and water ice. We seem to be here at just the right time: the Kuiper Belt might no longer exist in 100 million years or so, its objects ground to dust in collisions or wearing away as comets which melt incrementally on their repeated trips around the Sun.

Above: *These photographs show the movement of a moon around KBO OR10, the largest unnamed object in the solar system.*

Top: *An artist's impression of Quaoar, or KBO 2002 LM60. At the time of its discovery in 2002, Quaoar was the largest object found in the solar system since Pluto. At 1.6 billion km (1 billion miles) beyond Pluto, it's the most distant object in the solar system to be seen with a telescope.*

Below: *An artist's impression of Xena, KBO 2003 UB313, and its moon Gabriella. Xena is possibly slightly larger than Pluto. The Sun is in the upper left corner.*

Left: *An artist's impression of a KBO just half a kilometre (0.3 miles) across, detected by the Hubble Space Telescope but too small to photograph. This object has never been seen, but we know it's there because it passed in front of a star, disrupting the starlight that Hubble could pick up.*

NEW HORIZONS, ULTIMA THULE, 2019

The only KBO apart from Pluto to have been examined in detail is the snowman-shaped Ultima Thule, 6.5 billion km (4 billion miles) from the Sun. After leaving Pluto, New Horizons set a course for Ultima Thule, passing within 3,500 km (2,200 miles) in 2019. Ultima Thule is a contact binary – two bodies touching, probably now joined – with a total length of just 30 km (18 miles). Its almost circular orbit has not been perturbed since the start of the solar system, around 4.6 billion years ago. The brightest area is at the snowman's neck, where the two lobes join. The surface is pitted with craters or sinkholes, many as small as 0.7 km (0.5 miles) across. The smaller lobe is named Thule.

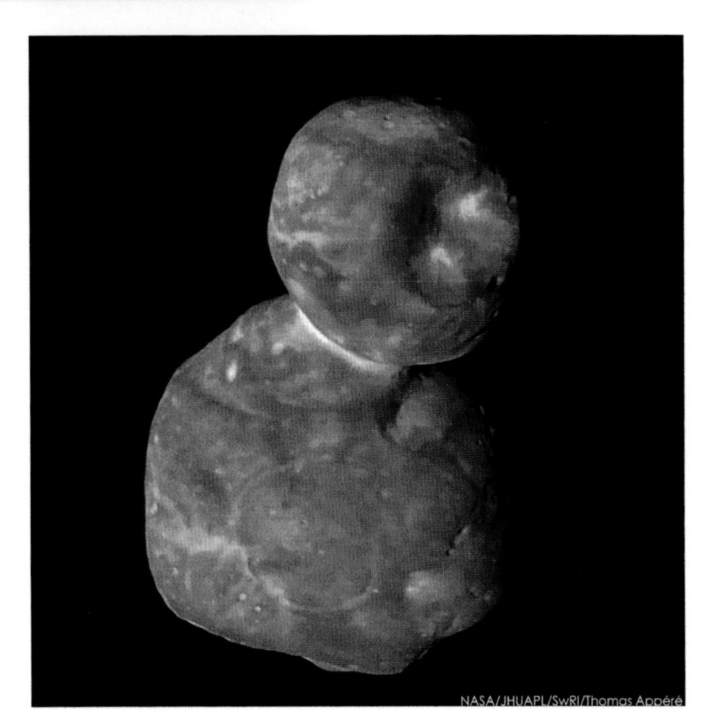

NASA/JHUAPL/SwRI/Thomas Appéré

A COMET'S TAIL

The Kuiper Belt is the source of the short-term comets that enter the inner solar system and loop around the Sun. When a KBO becomes a comet, it adopts an orbit which takes it closer to the Sun, leaving the Kuiper Belt. As the comet moves forward, dust released by the melting ice streams out behind it, forming a tail.

Comets from beyond the Kuiper Belt can have orbital periods of many hundreds, thousands or perhaps millions of years. The longest known is Comet West, with a return period of 250,000 years.

Left: *A Hubble Space Telescope image of the comet C/2017 K2, with the coma developing around it as it approaches the Sun. The comet originates in the Oort Cloud and has travelled for millions of years to reach its position in this photo, just beyond the orbit of Saturn. The coma is already 130,000 km (80,000 miles) across.*

ESA, PHILAE, COMET 67P/CHURYUMOV-GERASIMENKO, 2015

The ESA's lander Philae, carried by Rosetta to Comet 67P/Churyumov-Gerasimenko, was the first craft to achieve a soft landing on a comet. Photographs taken on Rosetta's approach clearly show the comet's shape, with two lobes. After Philae's landing, close-up images of the surface showed a rugged landscape of dark rock and ice.

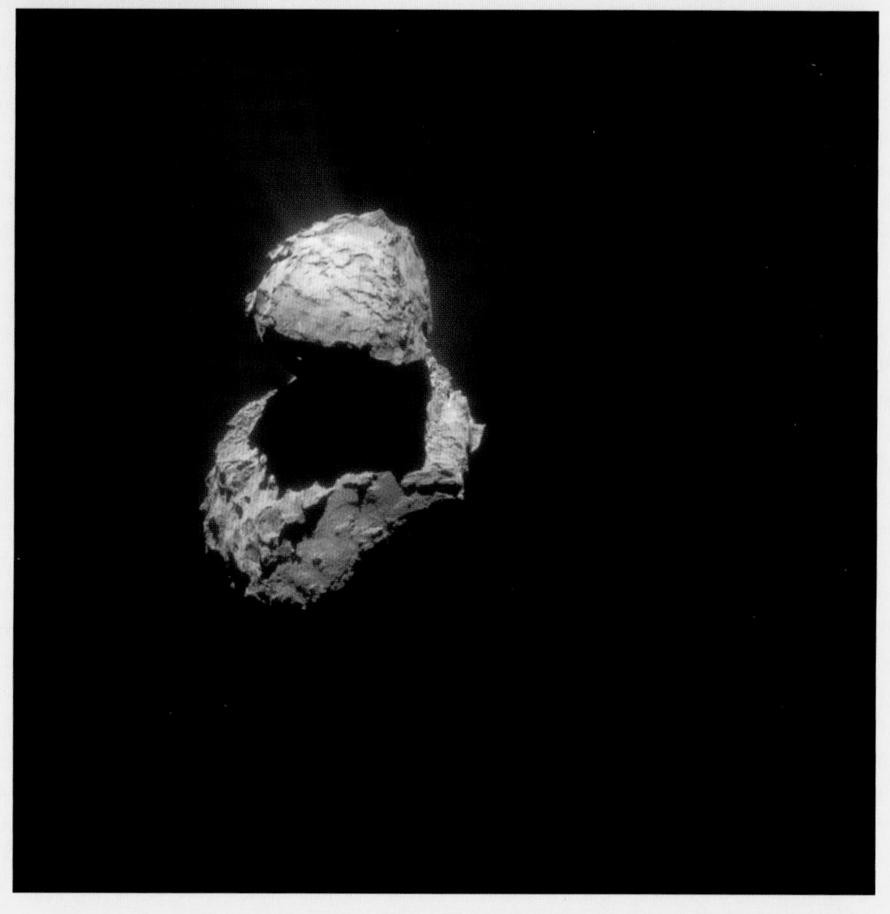

This photo taken at a distance of 124 km (77 miles) clearly shows the two lobes of the comet. A haze of dust from evaporating ice starts to form around it.

The lander Philae bounced when it landed, then rebounded and finally settled 1 km (0.5 mile) from the planned landing site, in an unknown region called Abydos.

The small lobe of the comet is to the left, with an area named the Hathor cliffs. The smooth neck of the comet, between the two lobes, is the Hapi region, strewn with boulders. This photo was taken from 30 km (19 miles) away and the region shown is 2.4 km (1.5 miles) across.

BEYOND THE BOUNDARY

We can only map our own solar system with any degree of confidence or level of detail, but its planets and other bodies represent just a tiny proportion of the billions of planets that must exist in the universe. Since the first exoplanets were discovered in the 1990s, astronomers have identified several thousand – and more are being found all the time.

Above: *Nineteen regions of the comet 67P/Churyumov-Gerasimenko have been named. Types of geological feature include pitted, smooth and dust-covered terrain, deep depressions and consolidated, rock-like surfaces.*

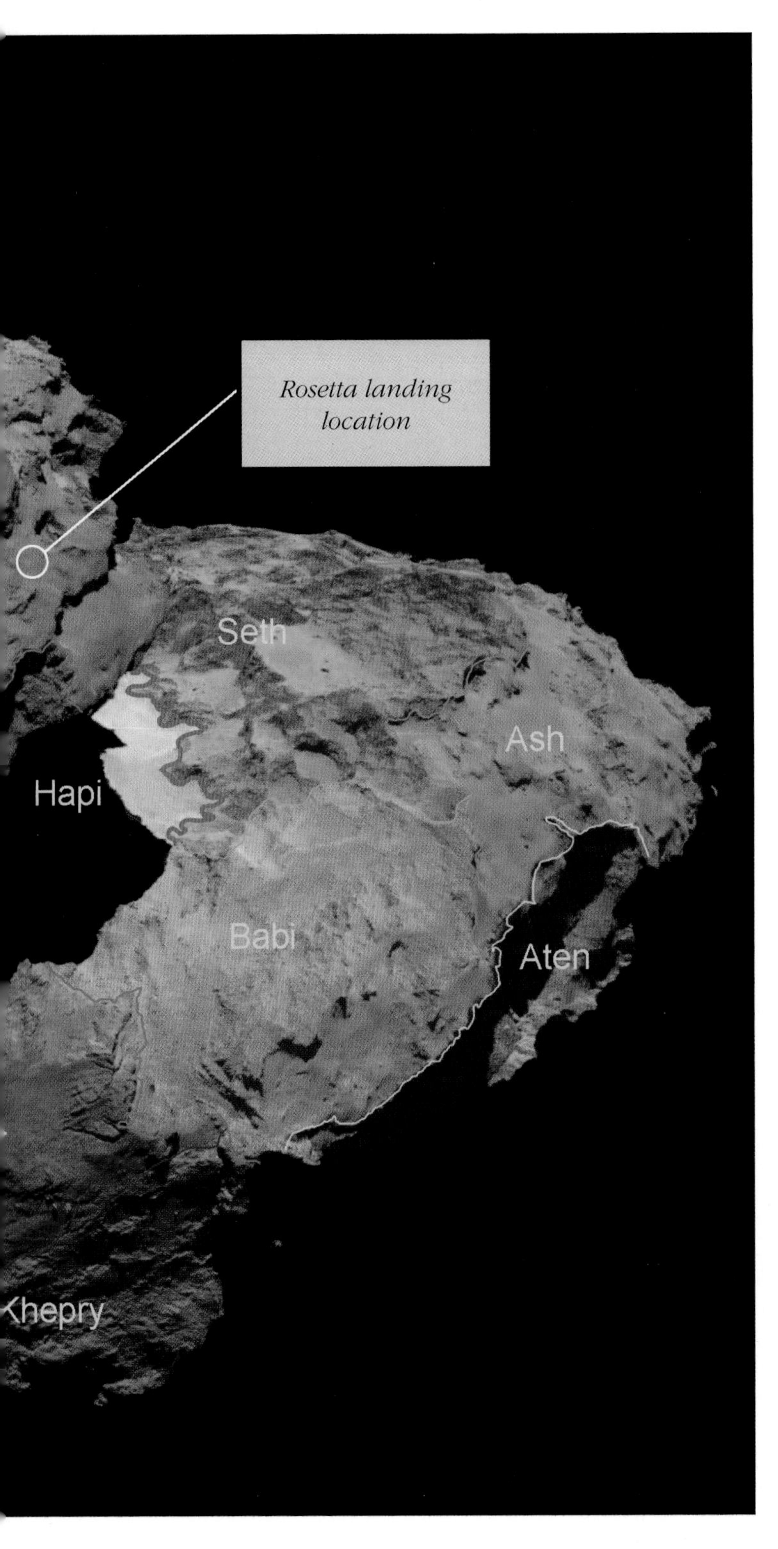

EUROPEAN SOUTHERN OBSERVATORY, ARTIST'S IMPRESSION OF 'OUMUAMUA, 2018

The first known visiting astronomical body from beyond the solar system arrived in 2017 and was named 'Oumuamua (below). It was spotted using the Pan-STARRS 1 telescope in Hawaii, which tracks near-Earth objects but saw 'Oumuamua a month after its closest approach to Earth. Although it was the first one identified, astronomers think such visitors from outside the solar system are not uncommon. It's likely that about two such encounters take place each year and one crashes into the Sun every 30 years or so. There are probably 10^{26} such objects in our galaxy, with a total mass 100 billion times that of Earth.

'Oumuamua is about 0.4 km (0.25 miles) long; it is dark red and made of rock and/or metal. It rotates every 7.3 hours. After looping around the Sun, it sped off at 158,360 kph (98,400 mph), heading back out of the solar system. Unlike a comet, it has no coma so is thought to be solid rock. 'Oumuamua has given us a glimpse of the kind of visitor that must wander through our solar system unnoticed on a regular basis. Similar asteroids that never approach the Sun but drift through the outer reaches of the solar system will also go unnoticed.

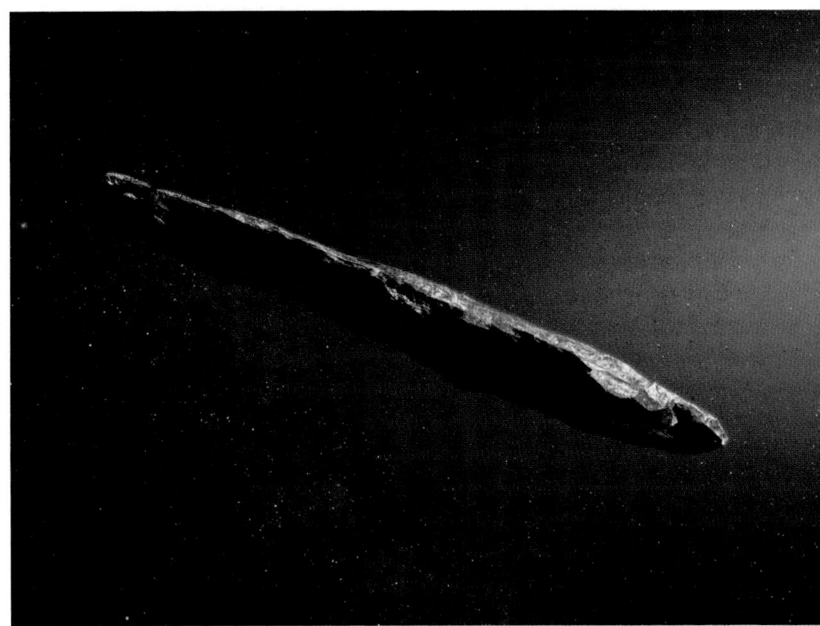

PLANETS BEYOND THE PLANETS

There is growing evidence that a further planet lies in the outer reaches of the solar system, causing the anomalous orbits of some Kuiper Belt Objects. If it exists, it's thought to be a 'super-Earth', up to ten times the mass of Earth and on an orbit around the Sun that takes 10,000–20,000 years to complete. It's suggested that there is another, tenth, planet too.

Far beyond the borders of the solar system we are discovering planets around other stars. These exoplanets have moved from science fiction to mainstream astronomy remarkably quickly. The first one was confirmed in 1992 and now thousands are known. Even the closest exoplanets are too distant to investigate with telescopes, but are often identified by the regular intermittent dimming of the host star as a planet passes in front of it. From studying the amount of dimming, astronomers can calculate the approximate size of an exoplanet, and from the interval between successive transits, they can calculate its orbit. Spectroscopy reveals some information about likely composition and temperature.

Astronomers have found gas giants, some of them far larger than our own gas planets and some of them hot (often called 'hot Jupiters'). They have found Earth-like rocky planets that orbit their stars at the optimal distance for liquid water to be present, and huge super-Earths. There are probably ocean worlds, too, with surfaces entirely covered with water.

Facing page: *A false-colour composite image from Hubble data shows the location of exoplanet Fomalhaut b. The planet follows a 2,000 year elliptical orbit around its star, passing through a crowded debris field. It will enter the debris field again in around twenty years, possibly experiencing collisions. The detail shows the position of the planet tracked over eight years. (The dark circle in the centre blocks out the light from the star so that the other objects are visible.)*

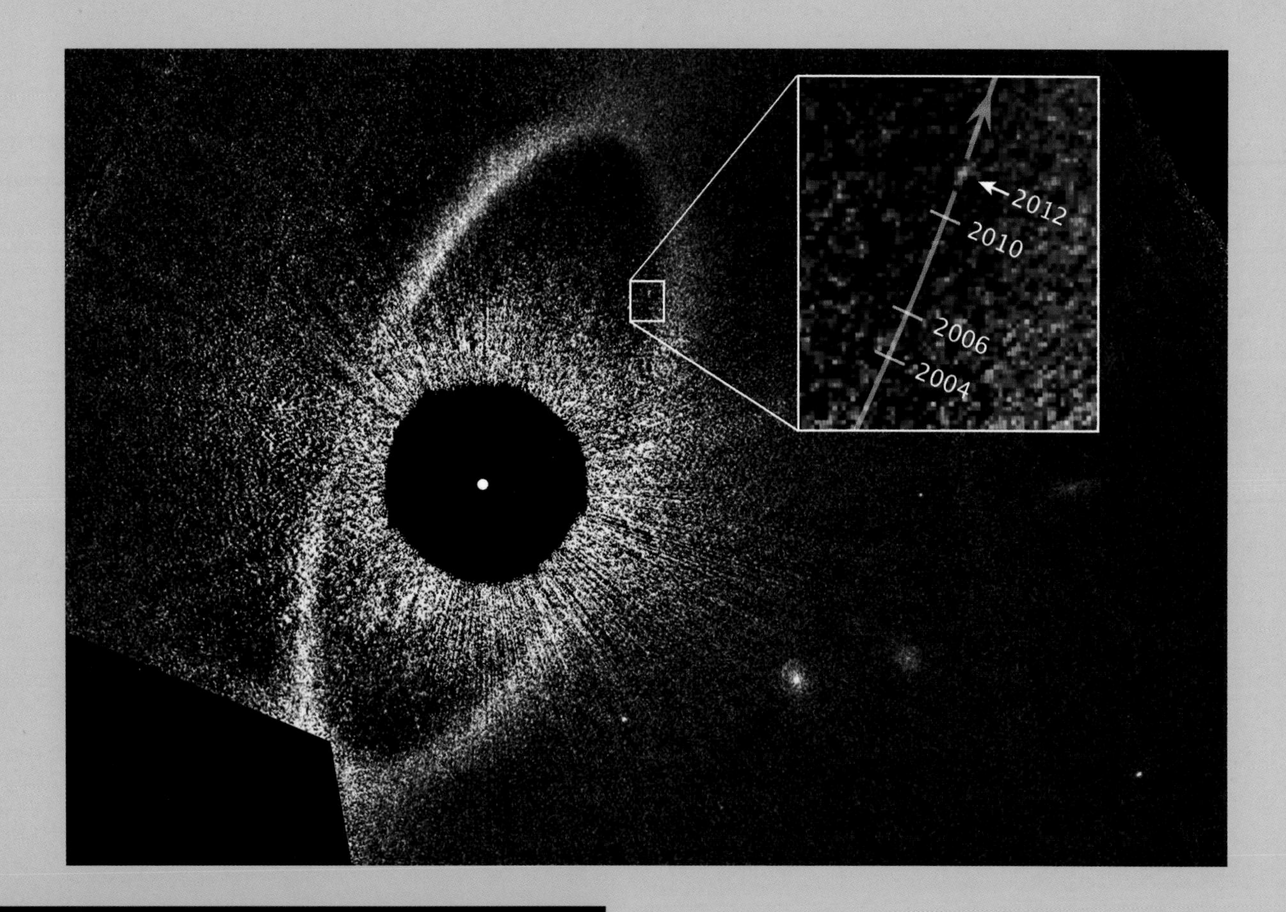

HOMELESS PLANETS

Because planets are defined as bodies in orbit around a star, the concept of 'rogue' planets at first seems odd. There are, however, possibly 200–400 billion orphan planets wandering through our galaxy. They may be planets thrown out of their solar system or they could have formed alone, accreting directly from interstellar dust and gas in the same way as stars. If they are not large enough to start nuclear fusion and become stars, they remain as wandering planets, hurtling unseen through empty space.

Left: *An artist's concept of the Trappist-1 system of exoplanets. The cool red dwarf star Trappist-1 has seven Earth-sized planets, one of which, Trappist-1e, is thought likely to have a global ocean of liquid water.*

INDEX

PICTURE CREDITS

akg-images: 12 (Pictures from History), 61; **Alamy Stock Photo:** 6–7 (BSIP SA); 10 (Science History Images); 11 top (Science History Images); 12–13 (Universal Images Group North America LLC); 20 bottom, 35, 36 top, 68, 72 top, 75 top, 77 top (Science History Images); 97 top (World History Archive); 148 top (Stocktrek Images, Inc.); 181 (Diego Barucco); **Bibliothèque Nationale de France:** 49, top; **BigStock:** 8–9, 17 (Peter Hermes Furian); 90 top; 103 bottom (Morphart); 122 top; 119 top; 146; 149; **Bridgeman Images:** 8 (British Library, London, UK; 24 top (Private Collection/photo c. Christie's Images); 48 (Bibliothèque Nationale, Paris, France); 72 bottom (University of St. Andrews, Scotland, UK); 174 top (Universal History Archive/UIG); 174 bottom (Musée de la Tapisserie, Bayeux, France); E. Kolmhofer, H. Raab; Johannes-Kepler-Observatory, Linz, Austria: 175 bottom; European Southern Observatory: 187 (M. Kornmesser); **European Space Agency:** 40; 42 top (MPS/DLR/IDA); 42 bottom (CNR-IASF, Rome, Italy, and Observatoire de Paris, France); 45; 55 x 3; 71, 84–5 (DLR/FU Berlin, G. Neukum); 89 bottom (DLR/FU Berlin); 140 (IPGP/Labex UnivEarthS/University Paris Diderot – C. Epitalon & C. Rodriguez); 141 (NASA/JPL/University of Arizona); 184 bottom, 185 top, 185 bottom (Rosetta/Navcam); 185 middle (Rosetta/Philae/CIVA); 186–7 (Rosetta/MPS for OSIRIS Team); **Getty Images:** 46–7; 58; 62; 103 top; 161 top (Corbis Historical); 162 (Hulton Archive/Stringer); **JAXA:** 43 top (ISAS/DARTS/Damio Boulc); **Library of Congress:** 77 bottom; **Mary Evans Picture Library:** 46; **NASA:** 14 (European Space Agency); 15 top (Johns Hopkins University Applied Physics Laboratory/Carnegie Institution of Washington); 15 bottom (Johns Hopkins University Applied Physics Laboratory/Southwest Research Institute/Alex Parker); 16 (JPL); 21 bottom (JPL/Malin Space Science Systems); 22 (JHUAPL/Carnegie Institution of Washington); 23 top (JPL); 23, bottom (JPL-Caltech/USGS); 24–25 (JPL-Caltech/UCLA/MPS/DLR/IDA); 26 (JHUAPL/Carnegie Institution of Washington); 29 top (JPL); 29 bottom; 30 x 2; 31 (JHUAPL/Carnegie Institution of Washington); 32 (Goddard Space Flight Center Scientific Visualization Studio/JHUAPL/Carnegie Institution of Washington); 33 top (JHUAPL/Carnegie Institution of Washington); 33 bottom (JHUAPL/Carnegie Institution of Washington/USGS/Arizona State University); 38 bottom (JPL-Caltech); 39 bottom (JPL); 41 x 2 (JPL/USGS); 54; 56 (NOAA); 57; 59 bottom right; 63 bottom; 66–7; 67 bottom (Colorado School of Mines/MIT/GSFC/Scientific Visualization Studio); 69 top; 69 bottom (Goddard Space Flight Center Scientific Visualization Studio); 70; 82–3; 86 (JPL/Arizona State University, R. Luk); 87 x 3 (JPL/ASU); 88 top (JPL-Caltech/Malin Space Science Systems/Texas A&M Univ. Phobos and Deimos images courtesy of NASA/JPL-Caltech/University of Arizona); 88 bottom; 89 top (HiRISE, MRO, LPL (U. Arizona); 90 bottom (JPL); 91 (JPL-Caltech/UCLA/MPS/DLR/IDA/ASI/INAF); 92 (JPL-Caltech/UCAL/MPS/DLR/IDA); 94–5; 97 bottom; 100 (JPL-Caltech/SwRI/MSSS); 101; 106 (Goddard Space Flight Center/JPL); 107 (JPL-Caltech/SwRI/MSSS/Gerald Eichstadt/Sean Doran); 108 top (JPL-Caltech/SwRI/MSSS/John Landino); 108 bottom; 109 (SwRI/MSSS/Gerald Eichstadt/Sean Doran); 110 x 2 (JPL-Caltech/SwRI/ASI/INAF/JIRAM); 111 (JPL-Caltech/SwRI); 112 bottom; 113 top (CXC/UCL/W. Dunn/JPL-Caltech/SwRI/MSSS); 113 bottom (JPL-Caltech/SwRI/MSSS/ Gerald Eichstadt/Sean Doran); 114 (ESA and the Hubble Heritage Team/STScl/AURA); 116; 117; 118 x 2 (JPL/University of Arizona); 119 bottom (JPL/University of Arizona); 122–3 (JPL/DLR); 123 right; 125 (JPL-Caltech/Space Science Institute/Kevin M. Gill); 126 (Erich Karkoschka/University of Arizona Lunar & Planetary Lab); 128 (JPL/SSI); 129, 130, 131 top (JPL/University of Colorado); 131 bottom, 132 (JPL-Caltech/Space Science Institute); 133 (JPL-Caltech/Keck); 134–5 top, 134 bottom, 135, 136–7 (JPL-Caltech/SSI/Hampton University); 138 bottom (JPL/SSI); 139; 142 (JPL-Caltech/ASI/USGS); 143 (JPL); 144 x 2, 145 top (ESA/JPL/SSI/Cassini Imaging Team); 145 middle and bottom (JPL/SSI); 147; 153 top (JPL); 153 bottom (JPL/STScl); 154–5 (ESA and M. Showalter/SETI Institute); 156 (JPL/Ted Stryk); 161 bottom (JPL); 166 top (JPL); 168 (ESA and J. Olmsted/STScl); 176 top (JPL-Caltech/UCLA/MPS/DLR/IDA); 177 x 2, 178 bottom, 179 x 2 (JHUAPL/SwRI); 180 top left (ESA and M. Brown/California Institute of Technology); 180 top right (Thierry Lombry); 182 top (G. Bacon/STScl and M. Brown/Caltech); 182 middle (ESA and A. Schaller/STScl); 182 bottom (ESA, C. Kiss/Konkoly Observatory and J. Stansberry/STScl); 183 top (ESA and G. Bacon/STScl); 183 bottom (JHUAPL/SwRI/Thomas Appéré); 184 top (ESA and D. Jewitt/UCLA); 188–9 (JPL-Caltech); 189 (ESA and P. Kalas/UCLA/Berkeley and SETI Institute); **National Geospatial-Intelligence Agency:** 49 bottom; **PLANET-C Project Team:** 43 bottom; **Science Photo Library:** 18–19 (Tim Brown); 27 (Detlev van Ravenswaay); 28 (Royal Astronomical Society); 34 (NASA); 51 (Philip Leat/Pete Bucktrout, British Antarctic Survey); 53 bottom (Science. Source); 73 bottom (Royal Astronomical Society); 75 bottom (Science, Industry and Business Library/New York Public Library); 93 bottom (Tim Brown); 98, 99 (NASA/JPL/Space Science Institute); 102 top; 102 bottom (Royal Astronomical Society); 104–5 (Damian Peach); 115; 122 left (Tim Brown); 150 (Mark Garlick); 152 (Library of Congress, Geography and Map Division); 154 bottom (NASA); 160 (NASA); 164 (Royal Astronomical Society); 165 top (NASA); 169 (Mark Garlick); 170–171 (Mark Garlick); 175 top (Tim Brown); 176 bottom (Mark Garlick); 180 bottom (John R. Foster); **United States Geological Survey:** 64–5; 78–9; 80–81; 81 bottom; 120–121 (USGS Astrogeology Science Center/Wheaton/NASA/JPL-Caltech); **Wikimedia:** 53 top (Eric Gaba).

Illustrations by David Woodroffe: 11 bottom, 157 top, 166 bottom, 172, 173 top, 178 top.